Numbers Tell Stories

Vaclav Smil is Distinguished Professor Emeritus at the University of Manitoba. He is the author of over forty books on topics including energy, environmental and population change, food production and nutrition, technical innovation, risk assessment, and public policy. His books for Viking/ Penguin, including the *New York Times* bestseller *How the World Really Works*, have been published in over thirty languages. No other living scientist has had more books (on a wide variety of topics) reviewed in *Nature*. A Fellow of the Royal Society of Canada, in 2010 he was named by *Foreign Policy* as one of the Top 100 Global Thinkers.

By the same author

China's Energy
Energy in the Developing World (edited with W. Knowland)
Energy Analysis and Agriculture (with P. Nachman and T. V. Long II)
Biomass Energies
The Bad Earth
Carbon Nitrogen Sulfur
Energy Food Environment
Energy in China's Modernization
General Energetics
China's Environmental Crisis
Global Ecology
Energy in World History
Cycles of Life
Energies
Feeding the World
Enriching the Earth
The Earth's Biosphere
Energy at the Crossroads
China's Past, China's Future
Creating the 20th Century
Transforming the 20th Century
Energy: A Beginner's Guide
Oil: A Beginner's Guide
Energy in Nature and Society
Global Catastrophes and Trends
Why America Is Not a New Rome
Energy Transitions
Energy Myths and Realities
Prime Movers of Globalization
Japan's Dietary Transition and Its Impacts (with K. Kobayashi)
Should We Eat Meat?
Harvesting the Biosphere
Made in the USA
Making the Modern World
Power Density
Natural Gas
Still the Iron Age
Energy Transitions (new edition)
Energy: A Beginner's Guide (new edition)
Energy and Civilization: A History
Oil: A Beginner's Guide (new edition)
Growth
Numbers Don't Lie
Grand Transitions
How the World Really Works
Invention and Innovation
Size
How to Feed the World
Creating and Transforming the Twentieth Century
Speed

Numbers Tell Stories

How Data Can Reveal the Surprising Truths about Our World

VACLAV SMIL

PENGUIN
VIKING

VIKING

UK | USA | Canada | Ireland | Australia
India | New Zealand | South Africa

Viking is part of the Penguin Random House group of companies
whose addresses can be found at global.penguinrandomhouse.com

Penguin Random House UK,
One Embassy Gardens, 8 Viaduct Gardens, London SW11 7BW

penguin.co.uk

Penguin Random House UK

First published 2026
001

Set in 13.4/16pt Bembo Book MT Pro
Typeset by Six Red Marbles UK, Thetford, Norfolk
Printed and bound in Great Britain by Clays Ltd, Elcograf S.p.A.

The authorized representative in the EEA is Penguin Random House Ireland,
Morrison Chambers, 32 Nassau Street, Dublin D02 YH68

A CIP catalogue record for this book is available from the British Library

ISBN: 978–0–241–75885–4

Penguin Random House is committed to a sustainable future
for our business, our readers and our planet. This book is made from
Forest Stewardship Council® certified paper.

FSC www.fsc.org
MIX
Paper
FSC® C018179

Contents

List of Illustrations

Introduction

We live in an age of noise—an endless churn of opinion, prediction, and concerns about how the world works. Yet beneath the headlines there are stubborn, often counterintuitive facts. The numbers can be stranger, more revealing, and more useful than the stories we tell. This book is an attempt to follow those numbers where they lead.

Take marriage. Weren't people in the past married off as children? In fact, in eighteenth-century France the median age at marriage was 29 for men and 27 for women—not so different from today. Take longevity. Should we look to Japan for the secret of long, healthy lives? Japan performs very well, but only one in five Japanese men and women retains full independence in very old age. Take waste. Modern electronics now contain more gold per ton of discarded waste than some gold ores—our rubbish is literally richer than some mineral resources. These facts open windows onto deeper truths about how we live, why we make the choices we do, and how we might change them. Other numbers have larger consequences. Cutting greenhouse gas emissions on anything like the recommended timetable would require resources on the scale of a major national mobilization—roughly 15–20 percent of the annual GDP in wealthy countries, comparable to the Allied effort in the Second World War.

Each section of this book takes a different angle—fertility and longevity, food and fuel, materials and machines, trade and work—and answers a single question: What do the facts really show? That the Japanese now get, on average, more food energy from milk than from fish; that the maximum power of explosives has grown by as many orders of magnitude as the number of components on a microchip; that the latest commercial jetliners do not fly any faster than the first designs of the 1950s. The aim is to replace assorted myths with measured realities and to make complexity legible again.

Every essay stands on its own and is fully referenced; together they form a portrait of a civilization that is ingenious, productive, and wasteful, dependent on concentrated energy and intricate supply chains, and still learning how to live with the consequences of its own strength. As always, every number has been checked and rechecked, but errors will remain—attentive readers should bring them to the attention of the publisher so they can be fixed. My words are not gospel, my numbers are not unimpeachable—but together they make it harder to be fooled.

Vaclav Smil — Winnipeg, 2025

The Shape of Humanity

A Marriage of Misconceptions and Reality

The history of marriage ages in England and Wales can be reliably reconstructed going back to the 16th century, when the average age at marriage was 29 for men and 26 for women. This may surprise those thinking of Shakespeare's Romeo and Juliet or of early-age royal betrothals (with many girls younger than 15 years), but the lives of most people had little to do with such literary and royal outliers: in Europe in the late medieval (1300–1500) and early modern (1500–1800) eras, couples contemplating marriage required years of savings to secure the expenses necessary to set up a new household.

That is why the median marriage in 18th-century France was no different from the English norm, at 29 years for men and 27 years for women. Of course, those averages hide plenty of early-age, and perfectly legal, marriages. In 1792, the new French revolutionary legislation set the minimum ages at just 13 years for females and 15 years for males, and in 1804 the Napoleonic Code raised it to 15 years old for females and 18 years old for males. As average incomes rose, the average ages of newlyweds drifted down slowly: in England, to 25 years for men and 24 years for women by the 1840s; in France, to 28 for men and 24 for women by 1880. But while the French means kept slowly falling, the English means went up again, peaked during the 1920s, and reached their lowest ages by the end of

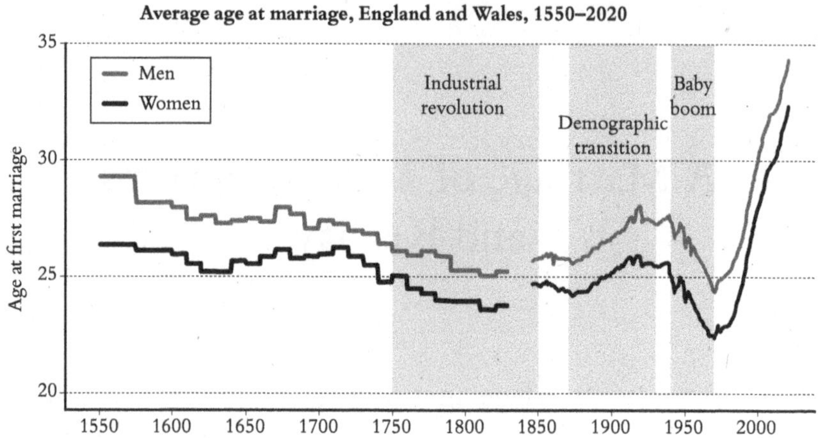

Average age at marriage

the baby boom era (1960s) before steeply rising to the recent means of nearly 35 years for men and 33 years for women.

This post-1960 rise has been the norm in every Western country. By the early 2020s the US averages had risen from 20 years for women and 22 years for men to 28.5 years for women and 30 years for men; in France they rose even higher, to 31 and 33 years. The reason for this widespread increase is a combination of many factors. More young people in general and more women in particular are pursuing higher education (one example of many: women now make up 50–70 percent of all medical school students in many Western countries, and with added internship and residency programs the majority of them will not finish their education until they are in their late twenties), and both sexes have higher expectations for marriage (above all, finding a similarly educated partner), leading to broken and postponed commitments. Also, many young people (and many more women in particular, given historical imbalances) are prioritizing their careers, while others are facing financial challenges—a reality most starkly illustrated by recalling the taken-for-granted ability with which young couples in the US and Canada were able to buy their first homes (now an unaffordable dream) two generations ago; now many couples are postponing marriage until they save enough for a down payment.

But there has been an even more consequential development regarding marriage: the decline of marriage rates. In many traditional societies the near-universal female marriage was the long-standing norm that persevered even into the second half of the 20th century. In China, non-marriage remained rare—only around 0.5 percent even among women born in 1960. But the post-Mao economic modernization, together with the one-child policy enacted in 1979 (which led to selective abortions of girls in a society that has traditionally valued males far above females, creating an enormous gender imbalance: by the early 2020s men outnumbered women by about 35 million), new educational opportunities for women, the high costs of marriage and child-rearing, and the rising practice of female hypergamy (educated women preferring to marry more-educated men with a higher economic status but finding such options ever more elusive) ended this millennia-long tradition.

In 2021, a Communist Youth League survey indicated that 44 percent of Chinese women aged 18–26 did not plan to marry, and, stunningly, in 2023 more than half of Chinese people between the ages of 25 and 29 remained unmarried—while the corresponding US numbers for that age cohort were nearly 50 percent for women and 60 percent for men. This translates to marriage rates (the annual number of marriages per 1,000 people) of 4.8 for China and 5.1 for the US, but these rates are far from the lowest globally, as marriages have become even less common in most of Europe (Germany 4.2, France and Spain 3.1) and Latin America (Mexico 3.3, Argentina 2.9, Venezuela 2.6—but Brazil's rate is still at 6.2!). Marriage rates remain high in Muslim countries, as well as in sub-Saharan Africa and in India where child marriage (girls younger than 18 years) remains a common problem.

In India, the Prohibition of Child Marriage Act (PCMA) of 2006 set the minimum ages at 18 years for women and 21 years for men, but the National Family Health Survey found in 2005–06 that 47 percent of women aged 20–24 were married before their 18th birthday, and in 2019–21 that rate was still 23 percent. Even in the villages belonging to the national capital region there were many cases of rich landowners marrying off their daughters at 15 or 16 years of age and spending "colossal amounts of money" on wedding celebrations and dowries. The PCMA amendment bill introduced in 2021 was to raise the minimum female age of marriage to 21 years, but it failed to pass.

The latest UNICEF data show that the shares of women aged 20–24 who were married before they reached 15 years of age are highest in Niger, the Central African Republic, and Chad (above 25 percent); among populous countries, in Bangladesh (16 percent) and Ethiopia (13 percent); and that in at least 50 countries around the world more than 25 percent of all marriages included women younger than 18 years of age. In sub-Saharan Africa, child marriages remain common not only in the poorest countries but also in Uganda and Kenya. The world is thus confronting two kinds of marriage age extremes—one increasingly late, one still too early, with no signs of any imminent resolution in either direction.

Fertility in Retreat

Total fertility rate (TFR) is the average number of children that the average woman is expected to have during her lifetime. The average replacement TFR (the number of children a woman must bear to replace herself and her partner) of a population is not 2.0 but 2.1, because not every child will survive to child-bearing age and because the normal sex ratio at birth slightly favors (105:100) boys. A population-wide TFR at 2.1 will keep the total count steady across generations, but in the past, with extraordinarily high infant mortalities (in premodern societies, one in five newborns died before their first birthday), TFRs had to be far higher—commonly at 5 or more—just to prevent population decline as fluctuating births and deaths kept the total numbers from rising rapidly. The highest recorded post-1800 fertility rates for large (national and regional) populations were between 7 and 8 in Africa, the Middle East, and Asia, and 5–6 in Europe—but in 1800 the French rate was already less than 4.5. Subsequent modernization (industrialization, urbanization, advances in hygiene, nutrition, and healthcare) lowered mortalities and ended the need for high TFRs to assure a desired number of surviving children.

British, US, and Italian TFRs declined from about 5 in 1870 to, respectively, 2.4, 3.1, and 3.5 by 1925. The post–Second World War baby boom temporarily raised the rates in Europe and North America to as much as 3.5, and subsequent declining fertilities have resulted from a concatenation of many changes,

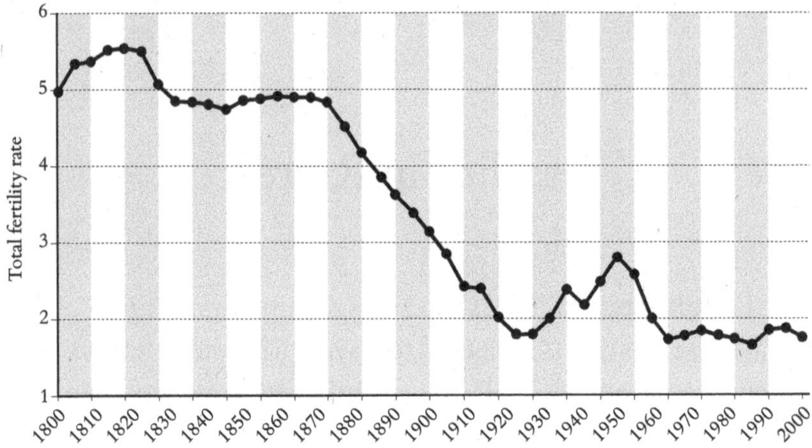

History of British total fertility rates

ranging from universal economic shifts to personal preferences. The continued urbanization and diversification of economies created many new jobs for women in office work, retail, and healthcare, and the unprecedented expansion of education at all levels (with record postsecondary enrollments) opened up many new jobs for women in teaching and administration. Changing career choices have seen a gradual rise of female participation in all academic disciplines: women now dominate North American medical and law schools, and only engineering remains predominantly male. As we have seen, increasing numbers of highly educated women (and, concurrently, of less-educated men) choose to postpone their marriages—and many women put their careers first, with about a third of them remaining unpartnered (see "A Marriage of Misconceptions and Reality," p. 3).

At the same time, living alone has become even more common among men: in 2022, 63 percent of American men aged 18–29 remained single! The cost of living and the cost of having children has been another factor lowering TFRs. The quest for material possessions and a higher quality of life led to preferential spending on goods (from houses and cars to clothes and electronics) and experiences (travel, attendance at cultural and sport events) even as the cost of child-rearing reached levels that became too high for many couples to contemplate having larger families (in North America, it now stands at no less than $300,000 until the age of 18, and more than $500,000 including postsecondary education). And, perhaps most fundamentally, it has become easy to realize the desire of having fewer children, as both contraception (the first hormonal pill was approved in 1960) and abortion (now legal in more than 70 countries) became readily available.

The resulting plunge in total fertility rates in nearly all affluent economies has yet to find its bottom. First, TFRs dipped just below the replacement level, fluctuating between 1.7 and 2.0, levels from which they could rather easily rebound to 2.1. By the 1990s, TFRs began to go just under 1.5 in some countries, a rate from which it would be much harder, but perhaps not impossible, to bounce back. In Europe, populous countries with TFRs below 1.5 now include Germany and Russia (1.4), Romania (1.3), Italy (1.2), Spain and Poland (1.1). How much lower can the rates go? By 2025 there were already several large populations with TFRs well below 1: Taiwan (24 million people, at 0.9) and South Korea (52 million people at 0.7), and even China is merely at 1.0, lower than Japan at 1.2.

Remarkably, declining TFRs soon spread to most medium- and low-income countries, with some of them going from traditionally high to near- or below-replacement levels in a single generation (in Iran from 6.4 in 1986 to 2.5 in 1999; in India from 4.1 in 1990 to 2.1 in 2025). Among the most populous nations, only Pakistan (3.5) and Nigeria (4.3) have very high TFRs, and by 2025 sub-Saharan Africa was the world's only region with high TFRs (4.2). The United Nations forecast that by 2050 the world's average TFR will be at the replacement level (2.1)—with regional rates ranging from 2.9 for sub-Saharan Africa to 1.47 for East and Southeast Asia, and with national extremes from 3.6 for Chad to 1.02 for South Korea. For Europe, North America, and Australia it is forecast at 1.55, slightly higher than the mean of 1.47 in 2025. Will affluent Western countries manage this slight recovery—or are such forecasts nothing but misplaced assumptions as TFRs keep on broadly retreating close to, and even below, 1?

We do not know, as all such long-term forecasts offer nothing but a choice of possible (and not necessarily plausible) outcomes. In any case, even relatively optimistic assumptions would result in steadily declining populations in Europe and East Asia. But there is one fertility-related forecast we can make with a high level of certainty. Unlike China's forced one-child policy, which was cruelly effective (and resulted in a huge sex imbalance, with a large surplus of males), government efforts to boost fertility will have only marginal effects. These campaigns appeal to patriotism and historic mission (now common in Russia, whose population is not only declining and aging but is also becoming less educated and is forced to cope with the consequences of the Ukrainian war) and offer financial incentives for additional births (in Russia, as well as in Singapore and South Korea). Baby bonuses would have to be far higher than the typical rates (all less than $10,000 per child) to start making a demographic difference.

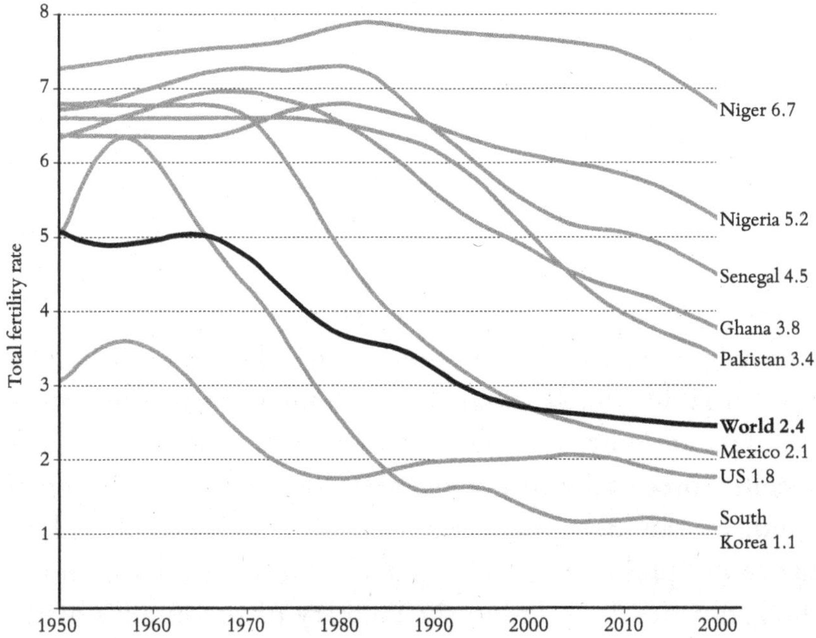

Total fertility rates around the world, 1950–2020

Living Forever

There is a lot of hype around some extremely wealthy individuals spending a fortune to optimize their health in the search for eternal life. Such gimmicks are uninteresting. The historical data are more compelling. While we cannot make any specific forecast about the eventual longevity of a newborn child, we can make interesting statistical inferences based on the recent probabilities of dying and surviving, based on the vital statistics collected in the country of the infant's birth. This information is packaged in life tables that are constructed by statisticians and demographers for a specific recent year (or for a short period of 2–3 years) by gathering data on the population's age distribution and on the number of deaths that took place within each age group. Because any specific change in life expectancy cannot be known in advance, they assume that these mortalities will hold, and use them to calculate mortality probabilities that are usually expressed (to avoid very small numbers) for a cohort of 100,000 individuals.

For example, in 2019 (before Covid-19 temporarily reversed the trend of slowly rising life expectancies) the Canadian mortality rate for boys during the first year of their lives was 0.00489—implying that about five male infants would die before the end of the year for every 1,000 males born in the country. This very low probability then declined even further (and substantially, by two orders of magnitude!) to reach its minima at 0.00007, or just seven deaths among 100,000 boys aged 7–9 years. These are the childhood years when death is the least likely (after surviving the riskier early years), but then its chances begin to rise, first slowly but always inexorably. At age 30, expected mortality in Canada during the year 2019 was 0.00101, or one death a year per 1,000 men; by 50 it was nearly three times as high (0.00287); by 75 it was 10 times higher than at 50 (0.02883), as nearly 29 men among 1,000 men of that age would die in a year; by 85, the total would be up to 82 per 1,000; and by 105 it would soar to 460 deaths per 1,000 remaining men, who would be facing a 50/50 chance of dying in that year.

This progress of mortalities can also be expressed in terms of surviving individuals. Of 100,000 Canadian boys born in 2019, 99,511 were expected to be alive at the end of the first year, 98,318 by age 30, 95,322 by 50, and 73,444 by age 75. Then the totals tumble, to 44,808 by 85 and to just 9,616 by 95. Obviously, remaining life expectancy shrinks with advancing age, from 80.11 years at birth to 51.22 at 30, 32.46 at 50, 6.51 at 85, and just 1.6 years at 105. When data have allowed, statisticians and demographers have been constructing life tables since the latter part of the 17th century (starting with John Graunt in 1662 and Edmond Halley in 1693), and for many countries you can now download their detailed annual and biannual sets—often going back decades—that are maintained and regularly recalculated by national statistical services.

Looking back at the progress of modern life expectancies makes for an impressive retrospective as improved hygiene and living conditions, better nutrition, and superior medical care (vaccinations, antibiotics, progress in the treatment of chronic diseases and cancers) have combined to maintain their steady—and cumulatively very substantial—increase. By 2025 the average combined life expectancy in affluent European countries had nearly doubled, from the mid-forties in 1900 to the early eighties. In Japan, now the country with the world's highest life expectancy, the overall increase has been even more pronounced, more than doubling the average life's length from 38.4 years in 1900 to 85.3 years in 2025, with the increments for 25-year periods being 4.9, 11.7, 13.6, 7.2, and 4.8 years—indicating the expected deceleration of longevity gains. How soon will these declining increments cease, signifying that we have reached the maximum human life expectancy? Or will scientific advances open the way for further gradual extensions of the human lifespan?

In the early years of the 21st century, several publications claimed that the latter development was already underway and that the continuing advances in medicine and genetics would soon translate into radical life extension. To begin with, these advances would extend the lifespans of infants born after the year 2000 to more than 100 years, and eventually they would go as far as the title of a 2004 book: *Fantastic Voyage: Live Long Enough to Live Forever*. A generation later, there is no convincing support for such claims. Since 1990, annual and decadal gains in average life expectancy have decelerated. In 2016 three researchers at the Albert Einstein College of Medicine argued that "the maximum lifespan of humans is fixed and subject to natural constraints," and in 2024 Jay Olshansky (at the University of Illinois in Chicago) and his coworkers analyzed aging in eight countries with the longest-lived populations, concluding that survival to 100 years is unlikely to surpass 15 percent for females and 5 percent for males, making any radical human life extension "implausible in this century": there is no easy way to reverse accumulated damage done on the cellular and tissue level.

And that may be a good thing, because increasing longevity

is a dubious quest unless it could be accompanied by the maintenance of good physical and mental quality of life, a combination that is very hard to maintain as our bodies age. Unfortunately, the statistics are clear: so far, our efforts have not been able to stave off widespread physical and mental decline among the oldest populations. What then is the value of longer lives when too many of them may be spent in demented states or warehoused in isolation in long-term care facilities? As we know, these realities are already having extensive personal, social, and financial impacts.

Again, the world's oldest-lived population illustrates these developments. In 2025 about 10 percent of elderly people in Japan lived with dementia, but projections indicate a significant rise of this share, perhaps up to nearly 29 percent by 2045. Even in Japan, where more very old people seem to be

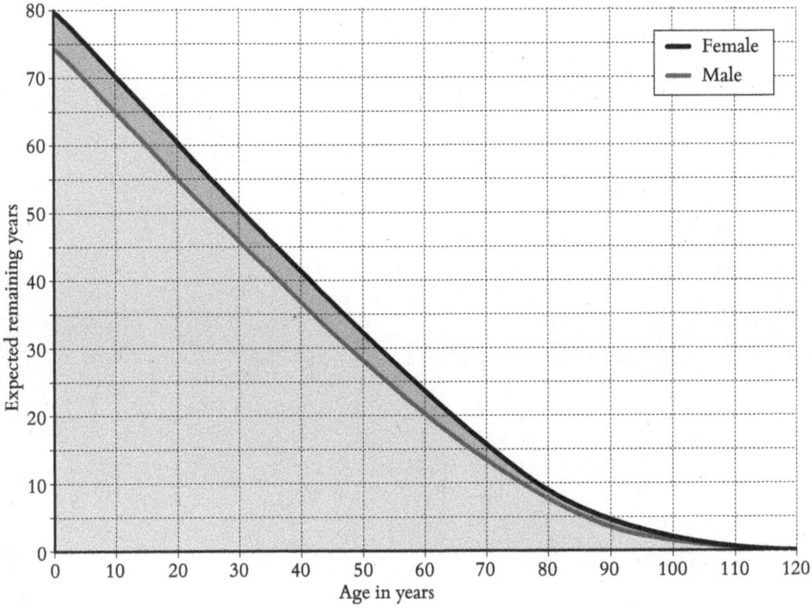

Remaining life expectancy in the US in the early 2020s

doing better than their counterparts in other nations, a study of Tokyo centenarians found that only 20 percent of them enjoyed physical and cognitive independence. And in 2023, in nearly two-thirds of homes in Japan where somebody over the age of 65 was receiving care, the caregiver was also an elderly person over 65—a high share that is bound to increase (with all its associated personal and familial stresses). That year the combination of aging and falling birth rates led a former prime minister to conclude that "Japan is standing on the verge of whether we can continue to function as a society." Sooner, rather than later, we will have to reconcile our wishes for extended lifespans with these sobering realities.

Too Much of a Good Thing

The mass-scale deployment of antibiotics (compounds that kill bacteria or prevent them from multiplying) has radically transformed the treatment of common bacterial infections. This fundamental advance has turned such infections from leading agents of mortality worldwide to overwhelmingly survivable medical complications. Before antibiotics became available, pneumonia (most often caused by *Streptococcus pneumoniae*) and tuberculosis (caused by *Mycobacterium tuberculosis*) were ubiquitous and indiscriminate death warrants—the first disease affecting all ages, the second killing mostly young adults in what should have been their most productive years.

Moreover, sepsis from wounds and surgeries (an injurious response to infections caused by many bacterial species, including *Escherichia coli*, *Salmonella enterica*, *Shigella dysenteriae*, *Pseudomonas aeruginosa*, *Neisseria meningitidis*, and *Staphylococcus aureus*) could commonly be fatal: the injury or the surgery could be minor, the sepsis would become deadly. As recognized by the World Health Organization, sepsis—because of its high mortality rate (25–30 percent, and 50–85 percent if it progresses to septic shock)—continues to be a major global health problem, with annual estimates put very conservatively at 30 million episodes and 6 million deaths per year.

The beginning of antibiotic development goes back to 1906, when Paul Ehrlich's group (at a private research institute in Frankfurt am Main) began to synthesize and test compounds that were derived from arsanilic acid against bacterial strains. In 1907, Alfred Bertheim prepared arsphenamine, and in 1909 Ehrlich and his assistant, Sahachiro Hata, discovered that compound killed (with no major side effects) the spirochaetes bacteria that cause syphilis. The first man-made antibacterial agent was approved in 1910 as a drug (Salvarsan) that became the standard treatment for syphilis for more than three decades. The next famous step came only on September 3, 1928, when Alexander Fleming, professor of bacteriology at London's St. Mary's Hospital, returned from vacation and noticed that in one of his petri dishes containing colonies of *Staphylococcus* (a bacterial genus causing sore throats and abscesses) a growing mold (later identified as *Penicillium notatum*) prevented bacterial growth.

But it took until 1945 to turn Fleming's accidental discovery of penicillin's antibacterial properties into a commercially produced drug used for treating ear, throat, urinary, and skin infections and for preventing rheumatic fever. Penicillin's introduction marked the beginning of the golden age of antibiotics, which followed a course common to so many inventions and innovations: after a slow start came a period of rapid ascent in the late 1940s and 1950s, as new antibiotics (including entirely new classes of these drugs) became available to treat many specific infections. Streptomycin (for serious bacterial infections) followed penicillin in 1946, chlortetracycline (for skin infections) and chloramphenicol (for eye and ear infections) in 1949, erythromycin (for pneumonia and skin infections) in 1952, vancomycin (for colitis) in 1958. Then the frequency of new antibiotic drug discoveries slowed down, with major additions including streptogramin (for skin infections) in 1973, rifaximin (for diarrhea) in 1987, linezolid (for pneumonia and skin infections) in 2000, and telavancin (for complicated skin infections) in 2009.

Antibiotics also made surgeries (from routine operations to organ transplants) safer as they minimized the risk of post-operative infections, enabled cancer chemotherapies that suppress the immune system and hence increase susceptibility to infections, and reduced maternal and neonatal deaths. Obviously, reduced morbidity and prevented mortality have large cumulative economic consequences, and these benefits are even higher when adding the impact of antibiotics in preventing infectious diseases among livestock—and hence lowering the costs of animal food production.

Estimates of the financial benefits of antibiotic treatments rest on many assumptions, but the cumulative gains can be large. In the US, the annual pre-antibiotic mortality rate for pneumonia was 70 per 100,000 persons, while for tuberculosis in 1940 it was 46 per 100,000: at these rates, without antibiotics, there would now be some 400,000 additional annual deaths in the US. Using $7.2 million as the average US value of a statistical life (this is an economic value used to quantify the benefit of avoiding a fatality), the savings would translate to nearly $3 trillion a year—and globally (at $1.3 million for an average worldwide value) it would amount to at least $12 trillion.

But concerns about new treatments increased as the number of new antibiotics approved for use in the US by the Federal Drug Administration declined from 16 between 1983 and 1987 to just two between 2008 and 2012; and although it bounced back to 14 between 2018 and 2022, concerns remain as only small numbers of new antibiotics were in development pipelines by the mid-2020s.

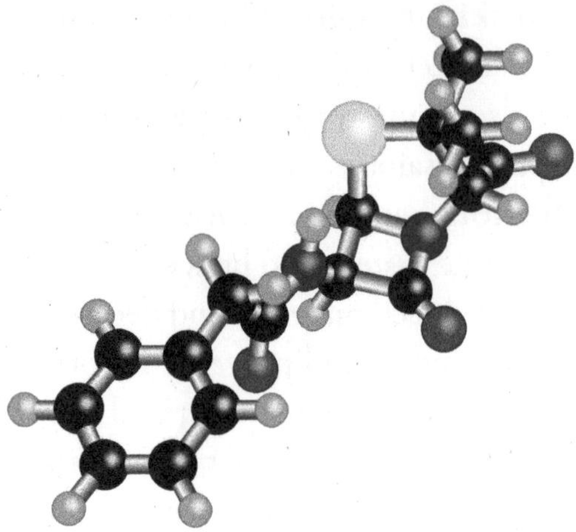

Penicillin molecule

New efforts are driven by the spread of antimicrobial resistance (AMR), which makes many long-established antibiotics ineffective and the underlying infections difficult or impossible to treat. This is an inevitable result of genetic adjustments accelerated by the overuse and misuse of antibiotics to control human and animal infections, made worse by the fact that in many countries (especially in Asia) antibiotics are available without prescription as if they were mere candies. A study of six Asian and African countries found that self-medication was widespread: the share of antibiotics dispensed without prescription was about 55 percent in Vietnam, about 45 percent in Bangladesh, and 36 percent in Ghana, and even in the EU only two-thirds of people in 14 countries who took part in a recent World Health Organization survey were able to say that their last course of antibiotics was obtained with a medical prescription.

The first global study of AMR impact estimated a total of 4.95 million deaths associated with the phenomenon, and 1.27 million deaths attributable to it in 2019. Managing AMR through correct usage and the introduction of new antibiotics remains a critical challenge for global civilization, especially as we now realize that nearly 60 percent of all infectious diseases affecting humanity worldwide have been, at some point, aggravated by climatic hazards whose impacts are not easily mitigated.

How Bad Was Covid-19?

Before Covid-19 began its global spread (during the fall of 2019 from Wuhan, although the world did not learn of it until January 2020), publications about pandemics were uncommon, limited to the writings of epidemiologists and historians—most commonly referring to the deadliest pandemic of the 20th century which began during the last year of the First World War. Unfortunately, the estimates of the global toll of that pandemic (wrongly but commonly referred to as the Spanish flu) remain far apart: we can trace its progress (but cannot pinpoint its place of origin), we have reconstructed the influenza A(H1N1) virus that was its cause, and we have some detailed studies of morbidity and mortality from European countries, the US, and Canada. The pandemic began early in 1918, and its third and final wave was spent only a year later. Published estimates of worldwide deaths range widely, from about 17 million to 100 million, with 50 million being perhaps the most likely total. If we divide that number by the 1.8 billion people that were alive at that time, we get a global mortality rate of about 2,800 per 100,000 or 2.8 percent.

Surprisingly, during the years of Covid-19 there were very few references to the two 20th-century pandemics for which we have more accurate mortality numbers. The first event was caused by the H2N2 virus that began to spread from China in February 1957, and the pandemic did not completely finish until the end of 1958. The second event was caused by the

H_3N_2 virus, which had also originated in China, in May 1968: its first wave peaked before the year's end, but in some countries the effects persisted until April 1970. The best reconstructions estimate that worldwide excess deaths—those presumably resulting from the pandemics—ranged, respectively, from 1.5 million to 4 million in the first event and from 1 million to 4 million in the second. Adjusting these mortalities for the changing global population number yields an excess mortality of about 52 per 100,000 (0.05 percent) from 1957 to 1958, and 30 per 100,000 (0.03 percent) from 1968 to 1970.

The first officially recognized pandemic of the 21st century was declared by the World Health Organization on June 11, 2009, and it was such a mild (and such an underreported) affair that less than 11 years later, when Covid-19 started, very few people (epidemiologists aside) could even recall it. There is a good reason for that. The H_1N_1 virus originated in January 2009 in Mexico and it stopped spreading before the end of the year, causing between 150,000 to 575,000 excess deaths worldwide. That was just 2.3–5.2 per 100,000 or a mere 0.002–0.005 percent, a toll whose lower rate was only twice the recurrent winter mortality caused by seasonal flu outbreaks (typically about 0.001 percent). In contrast, the latest pandemic, caused by the SARS-CoV-2 virus, was much more deadly.

According to the World Health Organization, by the end of 2024 there were some 750 million reported cases of Covid-19, and 7.1 million reported deaths. But perhaps the best global model of the disease, by the Institute for Health Metrics and Evaluation (at the University of Washington in Seattle), put excess mortality at 18.2 million between January 1, 2020 and December 31, 2021—that is 230 per 100,000, or 0.23 percent. That is slightly less than one-tenth of the 1918–19 toll, but about five times more than the excess mortality during the barely remembered 1957–58 pandemic and some 230 times more than during a typical seasonal flu, a difference large enough to make a deep mark in the memory.

Moreover, some countries saw much higher excess mortalities. Between 2020 and 2023 excess deaths in Europe were about 8 percent above the expected mortality, but the increase was 20 percent in Lithuania and 17 percent in Bulgaria (but less than 2 percent in Sweden), and during the peaks of the epidemic weekly mortalities were more than 100 percent above expectations! And unlike the 1918–19 pandemic that killed disproportionately large numbers of middle-aged people (with the peak excess mortality for both men and women in their thirties), Covid-19 killed mostly old people. The infection/fatality rate was 0.001 percent among children aged 5–9, 0.01 percent among adults aged 30–34, but it soared to nearly 10 percent among people older than 80.

The Covid-19 pandemic will not soon be forgotten. That is in a great contrast to the three preceding events. There are still hundreds of millions of people who lived through the 1957–58 and 1968–70 pandemics, and more than a billion people who are alive in the 2020s are old enough to remember the 2009 event—but very few of them have any distinct recollections of those officially recognized global emergencies. Is it just because those events had lower mortalities? Or because during those three pandemics we did not resort to any mass-scale economic lockdowns, did not enforce any long-lasting school closures, did not ban sports events, and did not cut domestic and international flights?

Covid-19 led to a substantial (70 percent to more than 90 percent) reduction in scheduled flights (during some hours the only commercial planes aloft were cargo planes carrying medical supplies from Asia) but, remarkably, during the earlier pandemics, aviation was marked by notable advances. On October 26, 1958—half a year after the end of the second pandemic wave in the West—PanAm inaugurated its Boeing 707 jet service to Europe by its New York–Paris flight. And the Boeing 747, the first wide-body jetliner, entered scheduled service months before the last wave of the H3N2 pandemic ended in March 1970.

Why were things so different back then? Was it because we had no fear-reinforcing 24/7 cable news, no Internet, and no incessant and instant case-and-death tickers on our electronic screens? Did we, as many people have argued, overreact during Covid-19 with blanket and prolonged restrictions that we had not imposed before? Almost certainly as far as school closures are concerned, but as somebody who was closing on 80 when

Type	**Global number of deaths for influenza pandemic**	
	Low Estimate	**High Estimate**
Spanish flu (1918–19)		
Patterson and Pyle (1991)	24.7 million	39.3 million
Johnson and Mueller (2002)	50 million	100 million
Spreeuwenberg et al. (2018)	17.4 million	
H2N2 virus (1957–58)		
Gatherer (2009)	1.5 million	
Michaelis et al. (2002)	2 million	4 million
H3N2 virus (1968–69)		
World Health Organization (2009)	1 million	4 million
Michaelis et al. (2009)	1 million	2 million
H1N1 virus (2009)		
Dawood et al. (2012)	150,000	575,000
Regular flu (in a year)		
World Health Organization (2017)	290,000	650,000

Comparing the pandemics

the pandemic began (and aware of the age-skewed mortality pattern), I appreciated the need for caution.

Or is it simply we ourselves who have changed, by subconsciously valuing recurrent but infrequent risks differently than in the past?

Can We Measure Happiness?

Your phone rings: it is the Gallup World Poll survey calling to find out how happy you are. If you decline, it might be because you are in a bad mood and the last thing you want to do is tell a distant voice about your imagined happiness (do not such reactions affect the survey's outcome?). If you agree to participate, you are told to "imagine a ladder with steps numbered from 0 at the bottom to 10 at the top. Suppose we say that the top of the ladder represents the best possible life for you and the bottom of the ladder represents the worst possible life for you. If the top step is 10 and the bottom step is 0, on which step of the ladder do you feel you personally stand at the present time?" Hadley Cantril, an American psychologist at Princeton University, came up with this concept for his public opinion studies, used it for the first time on the global scale in his book in 1965—and hence it became known as the Cantril Ladder.

About 1,000 subjective climbs up the ladder are obtained annually for each country. With 143 countries in the latest worldwide listing, that would work out to some 150,000 people or to less than 0.002 percent of the global population (in countries with insufficient phone ownership, the questioning is done directly). Rankings are calculated by taking three-year averages of these individual evaluations, adding them up, dividing them by national totals of respondents, and presenting the resulting numbers with the precision to the third decimal number. And so in 2021–23 the Swedes got to be the world's fourth-happiest people with the national score of 7.344, and Israelis with 7.341 got only fifth place in the global happiness race. You learned something: you never knew that an entire nation beats another nation because it is judged (just look at the quotient of the two numbers) 1.000408 times (that is, 0.04 percent) happier!

Who would have guessed that such a complex and subjective (and I would add elusive and variable) feeling as personal happiness could be gauged so precisely! Doubts may set in and, looking closer, you may start wondering why the two Nordic neighbors, the Finns (the world's happiest people) and the Swedes (number four)—the nations that are always floating at the top of global happiness charts—have suicide rates, respectively, 30 percent and 20 percent higher than Canadians, who come only 15th in the worldwide happiness rankings. And you may wonder even more when you see that the Finnish and Swedish suicide rates are higher than the rate in Japan, a country that is commonly seen as a suicide-prone nation and whose happiness index is, unhappily, only the world's 51st. Do you still endorse the ranking?

Of course, happiness to the third decimal point is just the result of adding up and averaging national responses—and perhaps the first rational reaction would be to round those numbers to the nearest 0.5. But no, that would not do: we must have clear ranking even if the differences are minuscule and forgettable or, as the science dealing with measurements would put it, statistically utterly insignificant. This is nothing unusual in modern societies, where the winner takes everything. Think of the gymnastics or figure skating competitions where somebody wins by a hundredth of a point, awarded after the scanning of rapidly moving and jumping figures by a small group of (often deeply biased) judges. For example, in the 2025 European Figure Skating Championship, the winner of the women's short program was 0.0725 percent ahead of the second-place skater.

And as far as the ranking of nations is concerned, we encounter the same dubious ordering phenomenon with the Human Development Index, which bases its global ranking on life expectancy at birth, expected years of schooling, and GDP per capita. In 2023–24, Switzerland was on top of the world with 0.967, but fifth-place Denmark was just 1.5 percent behind, and Canada in 18th place had an HDI just 3.3 percent lower. Given the inherent inaccuracies and dubious value of the constituent indicators making up the index (Do the expected years of school attendance tell us about the intensity and quality of imparted education? Does average GDP tell us about wealth distribution?), a much more sensible way than a definitive ranking based on the third decimal number would be to lump nations into deciles or quintiles (dividing the world into ten or five similar groups). Such an approach would also make sense for global happiness rankings.

These cautionary comments would be worth considering even if the measure were established by objective and reproducible procedures rather than by the subjective, sudden, inherently emotional, and near-instant challenge of placing one's life on an imaginary ladder. And that subjective placement (individual perceptions of happiness would have a considerable range even if objective circumstances were very similar) is also necessarily conditioned by cultural differences and expectations—including the self-reporting bias inculcated by specific national, ethnic, or religious circumstances. I will note just one important consideration pointed out by Boris Jardine (at Cambridge): when climbing the imaginary ladder that ends in perfect satisfaction, "the culturally loaded implication of striving and climbing, not to mention the idea of 'complete satisfaction,' are almost too obviously problematic to point out."

These, and many other inherent weaknesses of the now-standard approach to quantifying happiness, do not make the responses meaningless, but they certainly do not qualify them to be converted into rigid judgments suitable for accurate nationwide rankings differing by the third decimal number—and then to be used in a much-promoted annual publication to extoll certain populations as great winners of the global happiness race while relegating billions of people to dubiously self-assigned mental misery. The new happiness industry is now issuing an annual report claiming to be "the world's foremost publication on global wellbeing"—but I very much doubt that telephone polling based on the climbing of imaginary ladders can do that.

Happiness ladder in 2020

How We Feed Ourselves

Are We on the Way to a Vegan Planet?

Electronic media—those tireless sources of endless misinformation—spread many dubious claims without bothering to spend a few minutes checking the relevant facts. Have you realized that "much of the world is trending towards plant-based eating"? Moreover, "this global shift could be here to stay." So claims the Food Revolution Network. Other sources claim that, in just a few years, demand for meatless diets has seen 200 percent, 300 percent, or even 600 percent increases in some nations—as even burly sportsmen are becoming vegetarians eager to deliver unbelievable carrot-based performances. And the Vegan Society noted that in 2022 the sign-ups for their Veganuary campaign (people eat vegan during the month) was hitting record highs.

Does it mean that we are witnessing a new, and profoundly important, trend—and does it really apply to "much of the world"? Fortunately, this question is not difficult to answer, because we have reliable statistics going back decades, and half a century is surely plenty of time to discern a global trend. The easiest way—accessible to anybody with an Internet connection—is to check the meat supply data published annually by the Food and Agriculture Organization of the United Nations for all its (now about 200) member states and available also as continental and global aggregates. Two important qualifications: the FAO's data refer to total market supply (and not just to food that is eventually eaten); and they are expressed in terms of carcass weight (including bones and skin) rather than as edible meat (including offal, now eaten rarely in affluent countries but widely consumed elsewhere). Depending on the species, the share of edible/carcass weight is between 60 percent (beef) to 80 percent (poultry).

In 1975 the world produced 115 million tons of meat (including poultry), by the year 2000 it was about 230 million tons, and by 2025 it was about 380 million tons—more than tripling global output in half a century. Even after taking the intervening population growth into account (doubling from 4.1 to 8.2 billion), this amounts to an approximately 65 percent increase in average per capita terms (from 38 to 46 kilograms). This is as expected. Those providers of dubious news are unaware of an often-cited admonition made by Theodosius Dobzhansky, one of the 20th century's leading American geneticists and evolutionary biologists, who entitled his 1973 essay "Nothing in Biology Makes Sense Except in the Light of Evolution."

And our evolutionary heritage is clear: we, much like our primate ancestors, are omnivores, not herbivores; our guts are not those of leaf or seed eaters but of mixed-diet consumers because we have evolved to consume plants *and* animal foods. Chimpanzees, genetically the closest primates, deliberately hunt, kill, and share small monkeys, wild pigs, and tortoises—in some groups consuming 4–11 kilograms of meat per animal every year. That level of meat consumption by chimpanzees is on a par with many pre-industrial farming societies, where low crop yields precluded the feeding of domestic animal foods and where diets were dominated by plant foods. As soon as yields and disposable incomes began to rise—allowing higher shares of harvests to be fed to animals and spending less money on better diets (see "Food Expenditures," p. 51)—the populations of modernizing countries began to eat more meat, dairy products, and eggs: in Western economies this shift began during the second half of the 19th century; in most of Asia only after the 1950s.

Given the uneven stages of economic development, the weight of traditions and religious beliefs, population aging, and concerns about health, it is not surprising to see a wide range of national peculiarities. While the overall global demand for meat has been steadily rising, national trends show both increases and declines. Of all populous nations, only India, Bangladesh, Nigeria, and Ethiopia continue to eat very little meat. In 2025, average supply rates (in kilograms of carcass weight per year per capita) in India and Bangladesh were no higher than 5 kilograms, a bit less than in Ethiopia. But during the past 50 years, per capita meat supply in most of the world's populous countries has increased rather spectacularly: in Pakistan it has doubled (though is still less than 20 kilograms), in Turkey and in the Philippines it has more than doubled (in both countries to about 40 kilograms), it has tripled in Egypt (to more than 30 kilograms), Brazil's supply has more than tripled (to 100 kilograms), and in China it rose more than sevenfold (from only about 8 to just over 60 kilograms) and is now about 20 percent higher than in Japan!

Not surprisingly, meat supply has changed little in some traditionally highly carnivorous countries, including Canada, the UK, and Italy, and it has declined a bit in Germany, Denmark, and France: the avoidance of fatty red meat (driven by health concerns) by many younger consumers, higher intakes of seafood, and conversions of very small shares of populations to largely vegetarian (if not entirely vegan) diets have been among the factors behind these stagnations and retreats. This moderation of meat demand has been a welcome shift: the nutritional benefits of meat are not predicated on consuming it in very large amounts. But in other rich countries (including Australia, Sweden, and the US), demand for meat has reached new highs—and it has not led to any demonstrable ill-health effects.

Spain is the best example of this reality. Since Franco's death (in 1975), its average meat supply has more than doubled, peaking at nearly 120 kilograms in 2002; now, after declining to a still very high mean of about 100 kilograms, the country remains one of the world's top meat consumers besides the US, Australia, Brazil, and Canada. But even as meat demand was rising, the country's cardiovascular mortality was falling, and in 2019 (before Covid-19 affected average survival rates) Spain had, at 84 years, the EU's highest combined (male and female) life expectancy—notwithstanding all that *carne de cerdo asada*, *jamones*, and *chorizos* . . . Clearly, much of the world is not trending toward plant-based eating, and meat consumption does not shorten life expectancy—but lowered per capita meat consumption in highly carnivorous countries should be encouraged. But that's too sensible a conclusion to be popular.

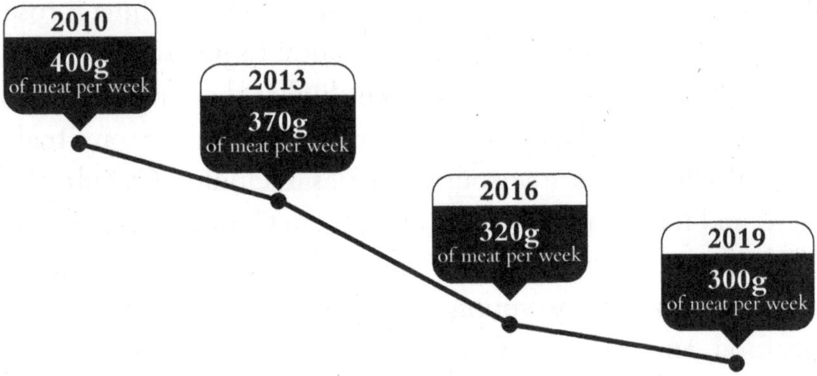

The French are becoming *petits consommateurs* of meat

Feeding Russia

During the two pre–First World War generations, Russian wheat cultivation expanded so rapidly that by 1903 the country was the world's largest exporter of this staple grain. Most of the harvest came from the Chernozem Region, with its naturally fertile, deep black soils extending north of the Black Sea and eastward to the Volga (that is, in today's Ukraine and south-central Russia), and then continuing east of the river. The abolishment of serfdom (in 1861), construction of railways, and significant expansion of cultivated areas were the leading factors enabling this growth. The liberation of the serfs led to higher agricultural productivity; the length of the Russian railways grew from just 1,000 kilometers in 1852 to 70,000 kilometers in 1913; and the area devoted to wheat cultivation doubled between the 1880s and 1910, with major extensions southeastward (the Krasnodar and Stavropol regions, north of the Caucasus) and eastward, advancing not only as far as Samara and Astrakhan on the Volga but even into western Siberia.

The US, benefitting from a similar (but westward, into the Great Plains) expansion of its wheat-growing area, led global wheat exports until 1902, then Russia took over between 1903 and the beginning of the First World War. The wheat price differential between Odessa and London was eliminated faster than that between New York and the UK, and by 1913 Russia exported 25 percent of the world's traded wheat as well as about 40 percent of rye and oats and 70 percent of barley. These exports were not without problems, as larger harvests came almost solely from expanding the land under cultivation rather than from raising yields (Russian wheat yields remained a third below the US mean), and Russian traders were involved in many disputes due to the poor quality of their deliveries (grain mixed with dirt and straw) and the careless extension of credit by Russian banks.

In Russia, the First World War was followed by years of civil war and the eventual establishment of the USSR in 1921. The country's agricultural recovery was reversed by the Stalinist collectivization of farming and by the deliberate starvation of Ukraine (Holodomor, 1932–33). Formerly the world's largest exporter of wheat, the USSR could produce barely enough grain to satisfy its growing domestic needs. And there were no radical improvements after Stalin's death in 1953. Khrushchev's interest in agriculture led to the mass-scale planting of low-yielding corn, and his ouster (in 1964) did not change this dismal farming performance.

The country's wheat supply was in such a parlous state that in 1972 the USSR turned to the US, and the Nixon administration mismanaged the massive trade deal (later known as the Great Grain Robbery) by selling the wheat at such low and taxpayer-subsidized prices that it led to soaring global costs of grain and to higher US inflation. Total Soviet wheat imports reached about 16 million tons in 1972, and after a brief period of fluctuation they began to rise in 1978, peaking at more than 28 million tons in 1984. The privatization of farming reversed this systemic failure of Soviet management: following the USSR's demise (in December 1991), wheat imports fell from 14.5 million tons in 1992 to just 5 million tons by 1999, and by 2001 Russia became a small net wheat exporter.

According to Western reporting, post-Soviet Russian affairs amounted to a litany of failures and miseries: these were undeniable, but there have been also some notable successes—and the return of Russian wheat exports to the world's top spot has been among the most consequential (albeit largely overlooked) achievements. Russian wheat exports rose from 10.6 million tons in 2005 to more than 18 million tons in 2008, more than doubled by 2020 and then reached a new record in the 2023–24 marketing year: just over 55 million tons of grain were exported. That was 24 percent above the previous year, and accounts for 26 percent of the global market share, a new record. Turkey, Egypt, Azerbaijan, Kazakhstan, and Pakistan were the largest buyers.

This time, unlike before the First World War, rising yields drove these rising exports as Russian wheat growers benefitted from an adequate supply of domestically produced fertilizers (Russia is the world's second-largest producer of ammonia, after China and ahead of India and the US) and domestic and imported pesticides and herbicides. Harvests averaged just 2 tons per hectare when the USSR collapsed; by 2022 they reached 3.5 t/ha, a 75 percent rise. Russia's production of agricultural machinery nearly quadrupled during the second decade of the 21st century, and the country now makes (and imports) large tractors and harvesters that are suitable for vast wheat-growing areas. Highly mechanized cultivation, timely harvests, and the development of new port capacities were key enablers of the rising exports: Novorossiysk on the northeastern shore of the Black Sea leads, with two terminals having an annual grain handling capacity of 11–12 million tons, and a further massive expansion of the Black Sea ports (to more than 100 million tons of grain capacity a year) is planned by 2035.

Moreover, Russia is now a prominent exporter of other foodstuffs: it is the world's largest exporter of frozen fish (ahead of China), the second-largest seller (after Ukraine) of sunflower oil, the fourth-largest provider of rye, and it has enormous potential to increase the output of every crop it grows. That is not surprising, given the country's size, variety of suitable growing environments, quality of its soils, and domestic capacity for producing agrochemicals and field machinery. The USSR was able to rival the US in some key technical areas (most notably nuclear weaponry, military aircraft, and crude oil and natural gas production), but even after 70 years of its existence it was unable to solve its agricultural underperformance. Post-USSR Russia has had its share of economic setbacks and political tragedies but its food production in general, and wheat cultivation in particular, has been a welcome example of success in a world where too many people (mostly in the Middle East and in Africa) still depend on imports of grain—and it also helps to explain how the newly aggressive, and much ostracized, Russia has maintained its global status.

Russia's largest wheat terminal in Novorossiysk on the Black Sea

Why Japanese Milk Is More Important than Its Fish

Milk in Japan has a very long history. Kōtoku, the 36th emperor (645–654 CE), was said to be so delighted with a gift of cow's milk that he called it a medicine that makes the human body better and arranged for milk production and the making of dairy products at the court. But soon afterwards came the first of many imperial prohibitions on eating the meat of all domestic animals—and that, combined with the necessity to secure food by intensive crop cultivation (leaving little room for pastures), worked against raising cattle. During the second half of the 16th century, when the Portuguese and other Europeans first reached Japan, they noted the presence of some small cattle used for plowing—but as Francesco Carletti, a Florentine merchant who visited the country in 1597, said, the people did not "drink their milk, for which indeed they feel as much disgust as we should for drinking blood."

In 1727, the Dutch—the only Westerners allowed to trade with closed Japan—gave three heads of white Brahman cattle to the shogun Yoshimune; by 1868, at the beginning of the Meiji Restoration that marked the end of Japan's more than two centuries of isolation, descendants of that gift numbered nearly 500 in a country of some 35 million people where hardly anybody drank animal milk. This long-lasting avoidance of milk-drinking brought high rates of lactose intolerance, a genetic trait caused by low levels—or even the total absence—of lactase, the enzyme that digests milk sugar, after early childhood. This complication has been documented by modern testing: the nationwide prevalence of lactose intolerance in Japan stands at about 73 percent.

High rates of lactase deficiency—in Japan and elsewhere in Asia, as well as in Africa and the Americas—have often been seen as insurmountable obstacles to incorporating dairy products into daily diets. Reality has been much more forgiving, however. Studies have shown that nearly three-quarters of Japanese are *not* prevented from consuming dairy products: in fact, post–Second World War Japan has seen an impressive increase in their per capita consumption. In most adults with lactose intolerance, lactase synthesis is not completely absent and this makes it possible to consume smaller volumes of milk without any discomfort: Japanese researchers found that only about 19 percent of Japanese adults cannot tolerate a daily consumption of up to 200 milliliters (slightly less than a cup) of milk, and when averaging just a cup of milk a day the annual per capita total would be about 80 liters a year.

Moreover, much larger amounts of fermented dairy products are easily digestible. Yogurts can have only half as much lactose as nonfermented milk; while unripened cheeses (such as cottage and ricotta) contain less than 30 percent of the lactose present in milk, Camembert has only a tiny share, and fully ripened well-aged hard cheeses (Cheddar, Parmigiano Reggiano) have no lactose left. Abstaining from dairy products was thus the consequence of historical and cultural circumstances, not a metabolically determined necessity. Once habits changed, Japan's historical statistics document a remarkable dietary shift.

After 1868, the westernizing government began to introduce

milk-drinking, but at the beginning of the 20th century only very few Japanese drank milk: average per capita consumption prorated to a negligible volume of less than 2 milliliters (less than half a teaspoon!) a day. In 1941, just before the beginning of what the Japanese call the Great Pacific War, the mean rose to a bit above 5 liters per capita, the equivalent of less than half a cup a week. Rapid consumption growth began during the 1950s, when milk was served regularly by the National School Lunch Program: by 1960 the national mean had more than quintupled to 20 liters per capita (less than 2 cups a week), and as the trend continued Japanese researchers attributed the unprecedented growth spurt of children that took place during the 15 years between 1960 and 1975 (12-year-old boys by as much as 7 centimeters, which was almost equal to the gain over the previous three decades) mainly to the increased consumption of milk (accompanied by more fish, meat, and egg protein).

By the year 2000 Japan had 1.3 million dairy animals (overwhelmingly in Hokkaido, the least-populated and northernmost island) and the average annual supply of dairy products (measured as fluid milk and including butter) had reached nearly 95 kilograms per capita, of which about 40 percent was drunk. A dietary survey found that only about 11 percent of all respondents did not drink any milk and 15 percent did not eat any yogurt. These rates came close to Japan's "peak dairy," and the first two decades of the 21st century saw only small gains amid the saturating demand. The latest statistics show the milk supply (excluding milk used to make butter) peaking at about 64 kilograms in 2018 and averaging just over 62 kilograms in 2022. This is a part of a larger trend, as the country's aging population has lowered the total daily per capita food supply to just below 2,700 kilocalories (kcal) a day (significantly lower

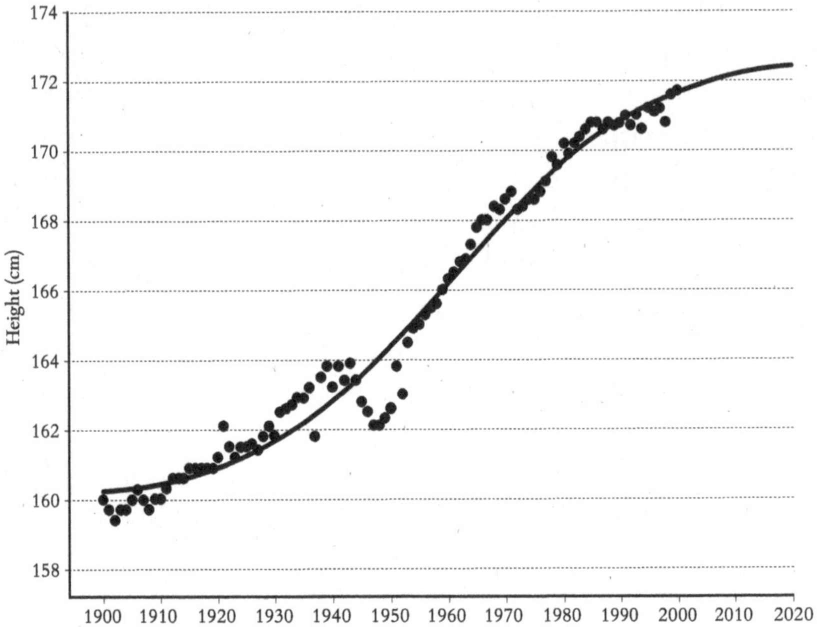

Average height of 18-year-old Japanese males, 1900–2020

than in China, at about 3,000 kcal; and in the EU or North America, where many countries average more than 3,300 kcal/day).

The latest supply data show, not surprisingly, that the Japanese still drink annually only about one-quarter of the milk annually consumed by the Dutch (even Americans do not match that superior rate!) but that their average annual consumption has come close (about 25 percent less) to that of India, the quintessential ancient milk-drinking civilization. Even more remarkable is the comparison in terms of average food energy supply. Although in decline, milled rice still dominates, averaging some 540 kcal per day per capita; but milk, with about 125 kcal/day/capita, is now ahead of seafood (about 95 kcal/day/capita). How many people would think that milk now provides more food energy in Japan than fish does? And how many Japanese would have envisaged that back in 1945, as the war-ravaged country was on the brink of famine?

Milk in a Japanese supermarket

The Madness of Blueberries from Peru

As a teenager I lived for several years near the Czech–Bavarian border in a place surrounded by forests, and picking small wild fruits—blueberries, blackberries, strawberries—during dewy mornings was one of the great summer delights. *Vaccinium myrtillum*, the European blueberry, a low shrub with tiny light-green leaves, yielded the most abundant harvests of the three. Alas, so briefly—and after a few forays of finger-staining picking came a year-long wait for a new crop. Less than 20 years later, after we arrived in Canada, so much had changed but fresh blueberries remained seasonal. Of course, they could still be picked in our great boreal forests, but the supermarket deliveries came mostly from farms in southern British Columbia, harvested and sold between August and September.

The plants from which they came were of the same genus as the tiny fruit I had picked in Europe, but from two different species. *Vaccinium angustifolium* (Canadian wild lowbush blueberry) was grown commercially in the east (mostly in Quebec). and *Vaccinium corymbosum* (North American highbush blueberry) in British Columbia. Highbush plants can grow taller than 3 meters, they have longer elliptical leaves, and they were harvested for millennia by Indigenous populations and then also by European newcomers.

In 1910, after a decade of studies aimed at domesticating the plant, Frederick Coville, a botanist working for the US Department of Agriculture, published a long report on *Experiments in Blueberry Culture*. The report was read by Elizabeth White, daughter of a cranberry farmer in Whitesbog, New Jersey, who offered Coville the family's farm for experimental plantings and solicited the locals to find the best *Vaccinium* bushes with large berries. Coville took the cuttings, cross-pollinated the plants, and by 1916 he and White had bushes with large and uniform berries ready for harvest.

Thus began the global march of domesticated commercial blueberries. For decades its reach was limited to several US states—as New Jersey was joined by Michigan, Minnesota, Oregon, and Washington in the north, and Georgia, North Carolina, Florida, and Texas in the south—as well as Canada (with harvests mostly from British Columbia). But by 2003, when New Jersey designated the highbush blueberry as the official state fruit, the blueberry cultivars had gained several major footholds outside of North America: in Australia since the early 1970s, in Poland since the late 1970s, in Chile since the late 1980s, and in Spain (Andalusia) and Mexico during the 1990s—with Chile and Mexico developing into major off-season suppliers for the North American and European markets. The subsequent expansion brought cultivated blueberries to every inhabited continent and multiplied their annual global production roughly tenfold. In 2010, only the US, Canada, Chile, and France produced annually more than 10,000 tons of blueberries; a decade later, 12 countries surpassed that rate; and in 2022 the global output reached nearly 1.9 million tons, with China being the largest producer (525,000 tons).

Peru was the second-largest producer (300,000 tons) and

the largest exporter (276,000 tons), accounting for a third of worldwide sales in that year. How has this tropical nation become the leading grower of berries from northern coniferous forests? Much like their wild ancestors, early commercial blueberry varieties needed 300–500 chilling hours with temperatures below 7°C, a condition unthinkable on the Peruvian coast—the country's main blueberry growing area, which is just 5–10 degrees south of the equator and where minimum temperatures never go below 10°C. Fortunately, some varieties of *Vaccinium corymbosum* proved adaptable—none more so than Biloxi, the shrub native to the southeast US that now dominates Peruvian harvests. And because the growers in the southern hemisphere (Peru, Chile, Argentina, South Africa, Zambia, Zimbabwe, Australia) have come to supply about 40 percent of global output, blueberries have ceased to be a seasonal product and are a commodity constantly available in supermarkets, be it in New York, Dubai, or Beijing. But as with every such convenience, year-round blueberries bring their environmental costs.

The extent of the land used for their cultivation is not the main concern (bushes prefer acidic soil not favored by most cultivated plants); rather, the distribution of the perishable crop leaves its mark and the berries would not look so perfect without regular applications of pesticides. Here are my scoping calculations approximating a key part of the first impact: Supermarket blueberries are sold in transparent hinged clamshells made of polyethylene terephthalate (the same plastic used for water and soft-drink bottles). When assuming that typical sales are in shells containing 300 grams of berries, each weighing 18 grams, the annual demand—for overwhelmingly single (buy, open, discard) use—would amount to about 100,000 tons of PET for 1.8 million tons of berries.

A Peruvian blueberry field

Calculations of the energy needed to transport the berries depend on several transport variables (shipping, trucking, air freight, container size, delivery distance, speed) that make it impossible to come up with representative global averages: Chilean berries go by boat, but nearly all the Peruvian berries have been recently flown to the US, a vastly more energy-intensive form of transportation.

As for the pesticides, in 2023 blueberries appeared, once again, on the infamous Dirty Dozen list of fruits and vegetables with the highest level of pesticide traces. Strawberries and spinach usually top the list, and blueberries ranked 11th, behind cherries and ahead of green beans. Moreover, to combat the spread of fruit flies and to reduce bacterial contamination with *Escherichia coli*, exported blueberries must be either cold-treated, fumigated with methyl bromide, or subjected to electron beam irradiation. Even so, storing the berries at 4°C for a week or two significantly reduces the antioxidant activity, largely negating one of the expected benefits of consuming the tiny fruit!

All in all, a very different product compared to the berries I picked in Bohemian forests more than half a century ago: no longer seasonal, much originating not only abroad but shipped or flown at high energy cost from different continents, with reduced antioxidant activity, containing undesirable residues, and packaged in throwaway plastic. *Bon appétit?*

Vaccinium angustifolium (wild lowbush blueberry)

The Universal Trend in the Cost of Food

People can be generally aware of certain social or economic relationships—but it takes a crisp formulation to make one of these realities into a generally applicable and much-cited law. Ernst Engel, a German economist and statistician, did that in 1857 in a paper on production and consumption conditions in the Kingdom of Saxony (the German Reich was established 14 years later). For those who cherish lengthy German sentences ending with a verb, the original reads: "*je ärmer eine Familie ist, einen desto grösseren Antheil von der Gesamtausgabe muss zur Beschaffung der Nahrung aufgewendet werden.*" In translation: "the poorer a family, the greater the proportion of its total expenditure that must be devoted to the provision of food."

This formulation has been so widely accepted that—130 years after its publication—Hendrik Houthakker, a leading American economist, could conclude that "of all empirical regularities observed in economic data, Engel's law is probably the best established; indeed it holds not only in the cross-section data where it was first observed, but has often been confirmed in time-series analysis as well." The law's ubiquity and solidity are not surprising. Of all existential necessities, eating must come first—and hence to secure adequate nutrition the poorest families in all societies will have to spend larger shares of disposable income than will better-off people.

And as incomes rise, the need for healthy food intake—be it in mass, energy, or nutrient terms—remains naturally limited (doubled income will not double overall food intake), and even with better (and more expensive) diets the relative expenditure on food will decline.

Average incomes began to improve significantly only during the 19th century, a time when many kinds of good statistical data regarding earnings and expenditures became available, and soon these were supplemented by regular national surveys of family expenditures. As a result, we can quantify the history of income share spent on food. During the mid-1970s, the Food and Agriculture Organization used food expenditure share as an indicator of poverty: a level above 59 percent indicated widespread food consumption shortages; with 50–59 percent, food intakes became adequate; countries with coefficients of 40–50 percent were considered relatively well-off; 30–40 percent meant being well-off; and values below 30 percent were found among the richest countries.

Using these standards, the US mean was in the relatively well-off range already in 1900, and half a century later it rose to the richest category. A reconstruction of income and spending published by the US Department of Labor shows that at the beginning of the 20th century the average US household spent 42.5 percent of its disposable income on food, compared to 23 percent on housing and 14 percent on apparel and services. By the end of the First World War the US share was a bit lower, at 38.2 percent, but in Boston it was as high as 44.7 percent—nearly three times as much as the families spent on clothes and services, and more than twice the monies spent on housing (including fuel, light, and furnishings).

During the economic crisis years of the mid-1930s, the food spending share in the US was still rather high, at about

34 percent, and New York's share was slightly higher at 36 percent. Only in 1950 did the nationwide share decline to just below 30 percent, slightly more than monies spent on housing. By the early 1970s it was just below 20 percent, and by the beginning of the 20th century the share of family income spent on food fell to 13 percent. A different food expenditure series published annually by the US Department of Agriculture includes food eaten at home as well as away from home, and it shows the share dropping to 9.97 percent in the year 2000, levelling off during the next decade, but by 2013 rising once again to just above 10 percent, while post-Covid inflation pushed it to 11.2 percent in 2023. An important secular shift hidden by this overall share is the rise in spending on food eaten away from home, from less than 30 percent in 1950 to more than half in recent years.

The best available international comparisons use a different metric: the share of consumer expenditure (including non-alcoholic beverages) on food eaten at home rather than on all food and beverage purchases. In 2022 this share was just 6.7 percent in the US, 8.5 percent in the UK, 9.5 percent in Canada, 11.8 percent in Germany, 13.6 percent in France, and 16 percent in Japan. In contrast, among low-income nations, food expenditures remained high. Nigeria, Africa's most populous nation and a chronically mismanaged economy unable to take advantage of its mineral wealth and plenty of farmland, was worst-off, with a share of 59 percent. Bangladesh, the poorest country of the Asian subcontinent, was at nearly 53 percent, Ethiopia at 40 percent, India at 32 percent, and China (whose share at the time of Mao's death in 1976 was as high as Nigeria's in the 2020s) still at 20 percent. But that is lower than both Argentina (23 percent), a major food exporter, and Mexico (26 percent). And Saudi Arabia, heavily dependent

Share of consumer food spending vs. value of total expenditures for selected countries, 2022

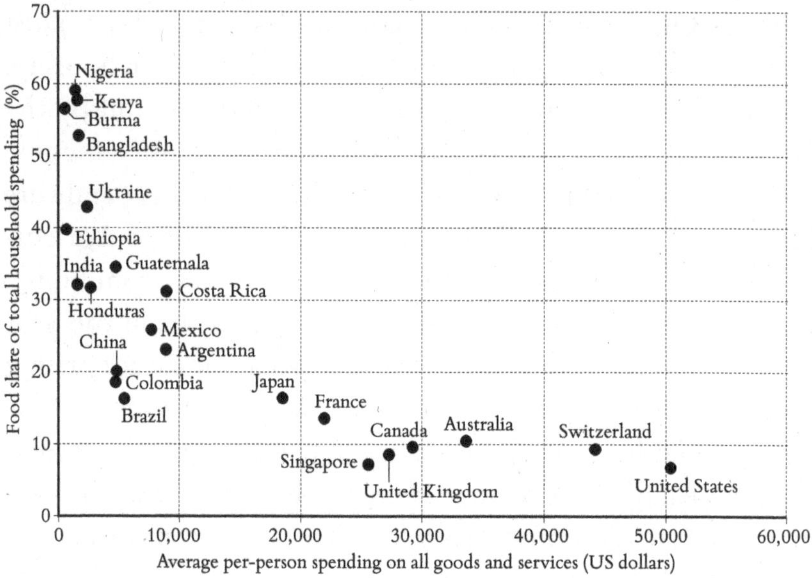

Food shares of household spending in 2022

on importing all foodstuffs, uses its wealth to keep the average share of food expenditure at about 20 percent, lower than in Russia (about 29 percent)—now the world's largest wheat and frozen fish exporter (see "Feeding Russia," p. 37).

Lower food expenditures have been an essential precondition of rising global affluence: obviously, our homes, clothes, material possessions, and leisure activities would look very different if we had not lowered that necessity from more than 60 percent (common in pre-industrial societies) to below 20 percent or even below 10 percent. But as food expenditures decline with rising affluence, specific income elasticities of food demand (that is, the percentage change in food consumption in response to a 1 percent change in income) differ. A comprehensive study of Engel's law in China shows (expectedly) that legumes (the hard-to-digest protein of the poor) and grains had the lowest income elasticity (their consumption rose only marginally as their prices declined), while fruit, sugar, and meat had the highest income elasticities, signifying a classic shift in dietary food intakes: richer families everywhere switch to higher quality and greater diversity in their daily food intake. And it costs a lower share of their income than it did for the families of their parents: Engel's law endures!

Powering the World

The Architecture of Air Pollution

Smokestacks—vertical pipes channeling the products of fuel combustion to higher elevations to reduce ground-level concentrations of solids and gases—became inevitable once the use of coal moved from heating houses and energizing small manufacturers to large boilers in factories. Many English artisanal workshops (forging iron, making bricks, tiles, or soap) relied on coal already during the 17th century, but only the invention of a more efficient steam engine (James Watt's original patent was granted in 1769), and its further improvements and expanded deployment in old and new industries, made it necessary to build taller chimneys to avoid releasing smoke near the ground and causing intolerable levels of air pollution.

By the 1840s, paintings and watercolors of English industrial cities (Manchester, Sheffield, Liverpool, Leeds) showed multitudes of tall chimneys belching out black smoke. But Scotland had the tallest structures: in 1842 Tennant's Stalk at Glasgow's large St. Rollox chemical factory became the world's tallest chimney, rising to nearly 133 meters; and 17 years later in the same city Joseph Townsend had an even taller structure (138.38 meters, requiring 1.4 million bricks) built at his chemical factory. Thousands of smokestacks were added around the world once coal-fueled electricity generation became an indispensable driver of modernity. Thomas Edison's first two small coal-fired electricity-generating plants began operating in the UK (in London at Holborn

Viaduct) and the US (in New York on Pearl Street) in 1882, and hundreds of new power plant chimneys followed before 1900. The expanding iron and steel industry (between 1870 and 1900 the global output of steel rose more than 50-fold!) added many more.

Smokestacks marked any late 19th-century cityscape. Claude Monet's most famous painting of the rising sun in the port of Le Havre in 1872—*Impression, soleil levant*, which gave the name to the entire school of late 19th-century painting—has ship masts and factory smokestacks as its vertical dominants. All these 19th- and early 20th-century smokestacks were brick-and-mortar structures, and without any emission controls the escaping gases were mixtures of carbon dioxide (CO_2, the primary product of combustion and the most important anthropogenic greenhouse gas) and carbon monoxide (CO, the result of incomplete combustion), as well as water vapor heavily contaminated with sulfur dioxide (SO_2, as nearly all coal contains traces of sulfur) and nitrogen oxides (NO and NO_2). The blackness of the smoke was due to particulate matter of different sizes, including the smallest particles (with diameters less than 2.5 micrometers) that cause atmospheric haze, irritate eyes, and penetrate further into lungs than larger particles.

As a result, when atmospheric conditions were right (no wind; temperature inversion), smokestack emissions generated high concentrations of airborne particulate matter and harmful chemicals near the ground, creating infamous smog episodes with very limited visibility and with widespread respiratory and eye irritation complaints and ailments (aggravated asthma, bronchitis). As the capacity of coal-fired electricity-generating plants grew after the Second World War, electrostatic precipitators became the norm at all new stations. They captured

Smokestacks in Manchester, c. 1852

more than 99 percent of particulate matter, but the escaping flue gases devoid of particulates were still laden with sulfur and nitrogen oxides—and so ever-higher stacks were designed to get the highest practicable dilution of these gases that were, starting in the late 1960s, implicated in causing the acid precipitation that was affecting fish and other biota, and damaging the growth of coniferous trees sometimes far downwind from major pollution sources.

The tallest stacks, all of them being reinforced concrete, were built at some of the largest coal-fired electricity-generating plants as well as at copper and lead smelters. The eventual US record height was reached in 1977 by a smokestack for Pennsylvania's Homer City coal-fired power plant (371 meters), while Canada's tallest stack, the Vale Superstack at a copper smelter in Ontario (380 meters), was finished in 1972. In 1987 Soviet engineers built the world's tallest stack (419.7 meters) serving the GRES-2 coal-fired power plant in Ekibastuz (now in Kazakhstan). Since the 1980s, as many cities in Europe and North America have become increasingly deindustrialized and turned into service centers, and as many utilities have switched to generating electricity from coal to natural gas, smokestacks—formerly skyline staples—have become useless eyesores. This has led to a thriving business of stack demolition that is done by carefully placed explosives that make the stacks either collapse instantly on their foundations or topple sideways. Some stacks have been gradually dismantled with the help of giant cranes and robotic machines.

But the age of the smokestack is far from over in China, where an unprecedented program of mass-scale coal-fired power plant construction (the country now has more than a thousand of them) and heavy industrialization (China is now by far the world's largest producer of steel, cement, and ammonia) began during the 1990s and led to the rise of thousands of new tall smokestacks. But by the time China embarked on this extensive modernization there was no need to build any tall superstacks, as the plants came equipped not only with high-efficiency electrostatic precipitators to capture all but a tiny share of escaping fly ash, but also with efficient desulfurization units: flue gases are led through wet scrubbers where a slurry of alkaline sorbent (usually limestone or lime) removes up to 95 percent of emitted sulfur oxides.

As a result, China's tallest stacks are not even among the top 100 tallest structures that still stand in the countries of the former USSR, across Europe, and in North America—but Chinese smokestacks, dispersing combustion gases from the annual burning of more than 4 billion tons of coal, are now by far the world's largest emitters of CO_2 and hence the dominant source of the most important anthropogenic greenhouse gas.

And India, now trying to replicate China's economic rise, will see more of them. Eventually the unfolding decarbonization of the global energy supply will end the smokestack era—but nobody can offer any confident forecasts of when none of these tall artifacts will be around.

Fuels vs. Batteries

The higher the energy density of transportation fuel (that is, the more energy contained in a standard volume), the better. The metric unit of the volumetric energy density of fuels is joules (the scientific unit of energy) per liter (that is, the volume of a cube with sides of 10 centimeters). When using a high-energy-density fuel, a moving contrivance (be it a road vehicle, a ship, or a plane) can have higher cargo-to-fuel ratio (and hence it will operate more profitably) as well as a longer range without refueling. This reality became of enormous commercial importance when the steam engine, the first fuel-powered prime mover, conquered land and ocean transportation. The era of steam locomotives and large steam-powered ocean liners began during the 19th century, when wood was still the world's dominant fuel, but neither locomotives nor ocean liners could be powered by wood-fired steam engines.

Dry soft wood (spruce, pine) has an energy density of only about 8 million joules (MJ) per liter, while good-quality bituminous coal has an energy density of at least 20 MJ/L or 2.5 times higher. Depending on the ship's design and engine's efficiency, transatlantic steamers had 30–50 percent of the vessel's volume dedicated to coal bunkers: using wood would have required allocating at least 75 percent of the ship's volume to stored fuel, making it impossible to carry goods and people profitably across the ocean. And with less-efficient engines, the crossing could not have been completed even if a steamer carried nothing else but wood!

Energy density also explains why ships switched from coal to diesel fuel once practical diesel engines became available: diesel oil packs 38 MJ/L, nearly twice the energy density of good bituminous coal, and its use has made mass-scale intercontinental shipping a quotidian reality. Similarly, diesel's high energy density enables long-distance trucking. A typical American 18-wheeler (semi: truck and trailer) weighs 15 tons and can carry 21 tons of cargo—but it needs only 150 kilograms of diesel fuel (a mere 1 percent of its own weight) for a trip of 500 kilometers. And energy density also explains why we cannot fly using any fuel containing less than about 30 MJ/L. The aviation kerosene that powers all modern jetliners contains 34 MJ/L, and when a Boeing 787 or Airbus A350 starts a long-haul flight across the ocean, 30–40 percent of its take-off weight is kerosene. Obviously, much lower energy densities would make such flights unprofitable.

But the unfolding quest for decarbonization aspires to displace fossil-carbon fuels and to use solar and wind conversions to electrify modern transportation. Where electricity is involved the preferred volumetric measure of stored energy density is watt-hours per liter (Wh/L). Air-dried softwood contains no more than about 2,500 Wh/L, good steam coal has around 5,500 Wh/L, aviation kerosene 9,400 Wh/L, and diesel oil about 10,500 Wh/L. How do electricity-storing batteries compare with the fuels that they are to displace in the carbon-free world? The first practical battery, invented by French physicist Gaston Planté in 1859, was a lead-acid cell, and its successors still dominate the storage of electricity needed in internal combustion vehicles (to start a car and to power lights, wipers, windows).

A lead-acid battery's energy density has gradually improved

from less than 60 Wh/L to about 90 Wh/L. The nickel-cadmium battery (invented by a Swedish engineer, Waldemar Jungner, in 1899) now has maxima above 150 Wh/L, and today's best commercial performers made in large factories are lithium-ion batteries, developed during the 1980s, with the first commercial version available in 1991. Currently, the best lithium-ion design commercially available in very large quantities is Panasonic's nickel-cobalt-aluminum (NCA) 2170 battery, widely used in electric cars. The single battery is a small cylinder 20.9 millimeters across and 70 millimeters long and weighs 66.9 grams; electric cars are each powered by assemblies of about 7,000 cells whose volumetric power density is 755 Wh/L, equal to 271 Wh/kg.

This means that a car traveling 500 kilometers without recharging will need about 410 kilograms of batteries (and associated parts) and its electric motor will weigh about 50 kilograms and transmission 40 kilograms, for a total of 500 kilograms. In contrast, an internal combustion engine will weigh 150 kilograms and the transmission and exhaust will

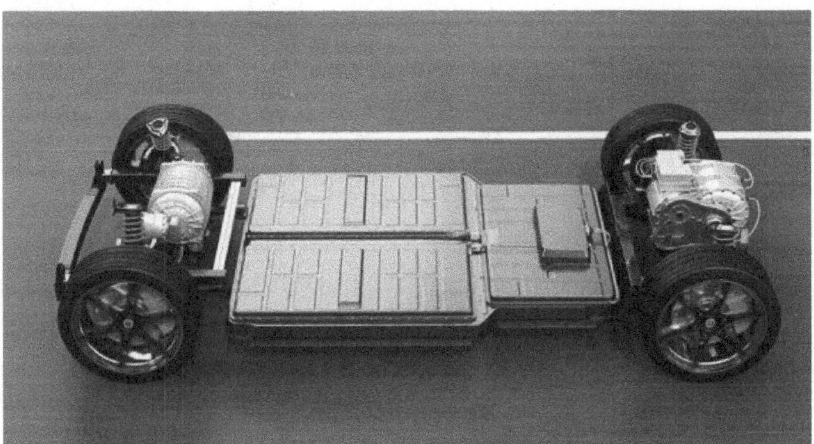

Battery pack in an electric car

add another 110 kilograms, and for the same trip it would need 25 kilograms of gasoline. The mass comparison—500 vs. 285 kilograms—is still very much in favor of a gasoline-fueled car. But battery density will improve. During the past 50 years the highest energy density of batteries available in large commercial volumes has roughly quintupled, from less than 150 Wh/L to more than 700 Wh/L. In 2020, Panasonic promised to boost it to about 850 Wh/L by 2025 (and to do this without expensive cobalt), and it is aiming eventually for a 1,000 Wh/L product.

Given the global interest in electricity storage, claims of new energy density records appear regularly, but Amprius Technologies of Fremont, California seems to be the leader with batteries rated as high as 1,150 Wh/L. But even if over the next 50 years we were, again, to quintuple that peak, then the top densities of about 5,500 Wh/L would be still no more than about half of diesel's volumetric energy density. What could we do with that? Battery-powered long-distance commercial flights would still be quite impossible, and even powering large container ships would work only in theory.

Total energy demand for a large container ship carrying about 20,000 steel boxes from China to Europe is about 64 gigawatt hours. To supply that demand, a ship powered by batteries with a power density of 1,150 Wh/L (or about 410 Wh/kg) would have to carry about 155,000 metric tons of them to go nonstop from Asia to Europe in 31 days. Those batteries alone would take up about 60 percent of the maximum cargo capacity, an economically ruinous proposition even when setting aside the difficulties involved in charging and operating the ship. The wait for super-batteries ready to power intercontinental shipping or the flight of wide-body planes may not be over even by 2070!

The Rise of Liquefied Natural Gas

Natural gas (methane, CH_4) is an excellent source of energy, but its low density (0.761 g/L compared to more than 850 g/L for commonly traded crude oil) precludes its waterborne intercontinental transport: a tanker loaded with crude oil would carry about 1,100 times more mass than the same shipping volume loaded with gas, making the latter option utterly uneconomical. The only way to reduce this physical disadvantage is to liquefy the gas. That is done by cooling it to −162°C and keeping it in insulated containers: this liquefied natural gas has a density 560 times that of gas at ambient temperature, and it contains more energy per unit mass—55 megajoules/kilogram (MJ/kg)—than diesel fuel (45 MJ/kg) or gasoline (46 MJ/kg).

But turning this understanding into a global business took more than a century. In principle, we have known how to liquefy gases ever since 1852, when James Joule and William Thomson (Lord Kelvin) discovered that highly compressed air forced through a porous nozzle cools slightly with every pressure drop, and that the repetition of this expansion results in a cooling cascade that can gradually bring any gas to its liquefaction point. Carl von Linde patented the first practical application of the Thomson–Joule effect in 1895, and the company he established remains one of the world's leaders in gas liquefaction.

In 1915 Godfrey Cabot patented the first design for shipping liquefied natural gas (LNG), but with coal extraction dominating both the world energy supply and demand, there was no incentive to commercialize this possibility. The next innovation came only in the early 1960s, motivated by the quest for cleaner air in England: unlike most types of coal, natural gas either contains no sulfur or the element can be easily removed before combustion, eliminating one of the most undesirable air pollutants. In 1961 the British Gas Board contracted for long-term imports of LNG from Algeria, and deliveries of gas from the Hassi R'Mel field via the port of Arzew to Canvey Island in the Thames Estuary began in 1964. Each of the two tankers (*Methane Princess* and *Methane Progress*) carried 27,400 cubic meters of the gas.

Discoveries of new natural gas fields in the Netherlands (Groningen) and in the North Sea, and soon afterwards the first Soviet natural gas pipelines into central Europe, dimmed the outlook for European LNG—but in 1969 two tankers began to export Alaskan LNG to Tokyo, while in 1971 the US began importing LNG to a regasification terminal in Massachusetts. This location, fewer than 4 kilometers from downtown Boston, raised concerns about the safety of LNG transport, but no terminal in the US, Europe, Australia, or Asia has ever had any loading/off-loading accident. After the collapse of the world oil price in 1986, Japan was left as the only major LNG importer—buying the gas under long-term contracts from Abu Dhabi, Indonesia, Malaysia, and Australia. South Korea and Taiwan were the next importers to sign long-term contracts (with Indonesia), and in 1996 Qatargas began shipping LNG from the North Field, the world's largest natural gas field, followed in 1999 by Trinidad and Tobago, and Nigeria. In the year 2000, 12 LNG exporters and a global fleet of about 100 tankers shipped the fuel to nine importing nations.

More fundamental changes came during the first two decades of the 21st century. Hydraulic fracturing (fracking—oil and gas extraction done by pumping pressurized mixtures of water and chemicals into underground deposits) led to an unprecedented decoupling of US oil and gas prices, and not just to a cancellation of plans for substantial LNG imports but to the US becoming by far the world's largest natural gas producer and leading LNG exporter (mainly through ports in Texas). And in 2007, Qatargas (now QatarEnergy) launched the first vessel of a new class of LNG tankers using membrane storage. Those LNG tanks can assume many shapes, as they are supported by the ship's double hull and made of several layers of sandwich-like insulation capable of carrying more than 250,000 cubic meters of LNG—or nearly ten times the volume possible in the early 1960s. By 2020 there were 19 LNG exporting nations, and a global fleet of 642 tankers shipped 371 million tons (nearly three times the 2000 total) to 43 nations.

Then, in February 2022, came the Russian full-scale invasion of Ukraine, and Europe, heavily dependent on

A Qatargas supertanker

Russian gas delivered by pipelines from Siberia, could not have coped with the sudden loss of most of that embargoed supply without emergency imports of LNG from the US, Nigeria, and the Middle East. By the beginning of 2025 the size of the global LNG tanker fleet had surpassed 800, the number of importing countries had risen to 55 (more than a quarter of the world's nations), and there were 22 exporters, ranging from Angola to Oman and from Equatorial Guinea to Norway. As a result, the global share of natural gas exported as LNG has been steadily increasing and in 2022 the volume of gas shipped by tankers surpassed exports via pipelines. And, in a most remarkable instance of safe operation (particularly given the early worries about the risk of LNG shipping), all of that without a single accident (loss of cargo, fire, explosion) in more than six decades of operation!

The history of global LNG illustrates perfectly how fundamental innovations need a concatenation of factors to succeed. We had the know-how to build the first LNG tankers right after the First World War, but in the absence of demand it would have been futile. Even half a century later, with crude oil globally ascendant, LNG remained a trading exception. Only early in the 21st century did all the preconditions for a massive global LNG industry—an adequate and diversified supply of affordable gas, a strategic preference for diversified supply, pre-existing national pipeline networks, high-capacity tankers, the rapid construction of liquefaction and offloading facilities—come together to make LNG an indispensable part of global energy use. Those who envisage a rapid (a decade or two) transition to a zero-carbon world should keep this lesson in mind: having PV cells and wind turbines is not enough!

The Global Energy Transition Will Profoundly Transform Modern Civilization (Slowly)

In his 1949 book *The Concept of Mind*, Gilbert Ryle, an English philosopher, introduced the term "category mistake." His first example of this common error is a visitor to Oxford who walks around and sees colleges and a splendid library and then asks, "But where is the university?" This category mistake is obvious: a university is an institution (and a very complex one), not a mere presence of buildings or books. Alas, too many people keep making even more egregious category mistakes and this leads them to many wrong conclusions. But few examples of category mistakes dominating the public discourse during the first quarter of the 21st century have been more important, and more consequential, than the way most studies and news reports see the global energy transition—the decarbonization of the world's energy supply that aims at the complete elimination of all fossil fuels.

The global energy transition is generally perceived as a discrete, well-bounded task that consists of replacing carbon fuels with non-carbon alternatives, and if that is not entirely possible by a given date, then to capture the remaining carbon emissions and sequester them underground, a so-called net-zero strategy. The urgency of the replacement (the longer it takes, the higher the average global temperature increase will be) has led some analysts and policymakers to call for solving it in the same way the United States dealt with the two notable challenges it faced during the past century: the rapid development of nuclear bombs during the early 1940s to preempt Hitler's Germany from getting them first, and winning the space race with the Soviet Union during the 1960s.

Simply put: Why not tackle decarbonization as an emergency must-do endeavor—as just another Manhattan Project, which culminated in the dropping of two nuclear bombs in August 1945, just three years after the effort's launch in August 1942? Or why not set the date for its accomplishment (2050 has been the most-cited year) and then spend whatever is required in order to reach it, much as was done with Project Apollo—which put two Americans on the Moon in August 1969, eight years and three months after President John F. Kennedy announced the goal in his special message to Congress on urgent national needs on May 25, 1961?

But to think about global decarbonization that way is to make an enormous category mistake. Difficult and costly as those two endeavors were, they were discrete efforts affecting only small parts of the American economy. Their costs were relatively modest when measured as shares of national economic product. And while underway, neither project affected the lives of American citizens in any profound way: indeed, the Manhattan Project was accomplished in such a secrecy that even Harry Truman, the country's vice president, was informed about it only after the president's death on April 12, 1945.

The decarbonization of the energy supply is the complete antithesis of those two programs. Ours is an overwhelmingly fossil-fueled civilization, and the size and complexity of the extensive super-system of fuel extraction, processing, distribution, storage, and conversion mean that its complete displacement will directly affect the lives of all people, the fate of all industries, and the performance of all critical subsystems ranging from growing food to long-distance transportation. Inevitably, its global costs will be commensurably high, far above the capacity of any nation effectively acting alone. A single example illustrates the magnitude of the challenge. By 2020 about 700 million tons of copper had been produced over the course of human history, but twice as much new copper will be needed to reach net-zero by 2050, mainly for replacing all internal combustion vehicles with electric drive systems (an electric car contains more than three times as much copper as a gasoline car) and for the massive electrification of industries.

As for the costs, by the time the Manhattan Project ended in 1946, the country had spent nearly $2 billion on it, or about $33 billion in 2025 monies. Divided over the three years, that

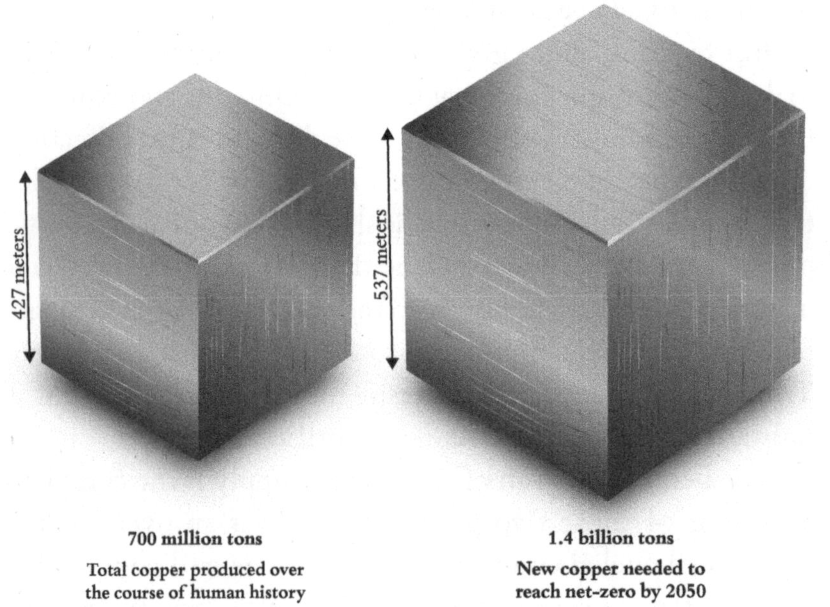

Copper required for the global energy transition

was about 1 percent of 1945's GDP per year. When Project Apollo ended in 1973, it had cost about $26 billion, or close to $200 billion in 2022 monies; over 12 years this worked out at about 0.2 percent of the country's GDP in 1969. Of course, nobody can provide a reliable account of the eventual cost of the global energy transition, because we do not know what will be the ultimate composition of the new primary energy supply—what shares will come from converting natural renewable flows, from using them to produce hydrogen or synthetic fuels, from relying on nuclear fission (and, as some hope, on fusion), or from mastering as-yet-unknown options.

But a recent attempt to estimate such costs confirms the magnitude of the category mistake propagated by those who call for equivalents of crash programs. The McKinsey Global Institute put the cost at $275 trillion between 2021 and 2050—roughly $9.2 trillion a year—compared to 2021's global economic product of $94 trillion. This is, as all long-range forecasts are, almost certainly incomplete and most likely a highly conservative estimate, but implies spending annually an equivalent of some 10 percent of today's world economic product to accomplish the task. And because the world's low-income countries could not carry such burdens, they have repeatedly demanded that affluent nations must finance most of their transition costs: that means that rich countries would have to devote on the order of 15–20 percent of their annual economic product to the task, shares whose magnitude is comparable only to resources that the Allies spent to win the Second World War. Are the governments and citizens of the EU, US, Canada, Japan, Australia, Britain and the Gulf oil states ready for such a decades-long commitment?

Microchips Have Grown Rapidly, but Something Has Exploded Even Faster

The relentless gains in crowding components onto a microchip—mostly transistors, resistors, and capacitors etched onto semiconducting silicon—have served as the ultimate, and widely used, example of the unsurpassed pace of modern innovation (See "Silicon," p. 92). This rise has been following Moore's law, named after Gordon Moore, one of the founders of Intel, who predicted in 1965 the doubling of components on an integrated circuit every year, and later adjusted that rate to two years for new microchips. The world's first microprocessor—Intel's 4004, released in 1971—had 2,300 transistors crowded on a silicon wafer, and half a century later the highest count has surpassed 5 trillion, a performance gain of nine orders of magnitude (2.3 billion times, to be exact). But these stunning gains in microelectronics have no counterpart in the history of other technical advances—be it the fastest speed of travel, growth of the tallest buildings, or gains in shipping capacities—as all non-electronic advances have seen only relatively modest gains limited by structural constraints or by the performances of prime movers (engines, turbines, motors).

During the entire 20th century, maximum travel speeds rose fewer than 10 times—from about 100 km/h for express trains in 1900 to 900 km/h for cruising jetliners that have been in uninterrupted commercial service since 1958 (but whose speeds have not shown any subsequent gain except for the short-lived and uneconomical supersonic Concorde). Skyscrapers became less than four times taller (119 meters for New York's Park Row Building completed in 1899 vs. 452 meters for Kuala Lumpur's Petronas Towers finished in 1998). And while the typical large ocean freighters could carry more than 10,000 tons in 1900, the largest size of the late 20th-century crude oil tankers surpassed 500,000 tons, for less than a 60-fold gain. But there is one accomplishment—unfortunately the most dangerous and the most destructive of all classes of inventions—that has seen not only even higher gains than the performance of solid-state electronics, but has done so in a much shorter time: the destructive power of post–Second World War explosives.

The history of modern explosives begins in the early 1860s with trinitrotoluene (TNT) and dynamite, and RDX (Royal Demolition Explosive) was invented in 1899. But in recent decades some of the deadliest terrorist attacks have been carried out using a mixture of commonly available fertilizer (ammonium nitrate, 94 percent) and fuel oil (6 percent), commonly known as ANFO. Based on peak pressure, the ANFO combination is equivalent to about 90 percent of the same mass of TNT—and hundreds of kilograms of it can be used to make very powerful truck or car bombs, such as those used by Hezbollah in 1983 to kill 241 US soldiers (mostly marines) in their Beirut barracks, and by Timothy McVeigh in the 1995 bombing that killed 169 people at the Alfred P. Murrah Federal Building in Oklahoma City.

During the Second World War, explosive power rained down by the Allied long-range bombers on European and

An ANFO car bomb was detonated in Oklahoma City in 1995

Japanese cities was done by the mass-scale dropping of incendiary bombs that burned Tokyo (on March 10, 1945) and other, largely wooden, Japanese cities, and destroyed Hamburg (July 1943) and Dresden (February 1945). By the war's end, the most powerful explosive weapon was the Nazi V-2 rocket designed by Wernher von Braun, a future American citizen whose Saturn rocket put *Apollo 11* on the Moon in 1969. The V-2 carried 910 kilograms of Amatol and had an explosive energy of about 3.8 gigajoules (GJ, 10^9 joules). For comparison, 1 ton of TNT is equal to about 4.2 GJ.

And then came an entirely new class of vastly more powerful explosives—those exploiting nuclear fission and fusion. The first fission bombs were developed by the Manhattan project (1942–45) and the best reconstruction of the power of the Hiroshima bomb (dropped on August 6, 1945) credits it with releasing 63 terajoules (TJ, 10^{12} joules) of energy: half of it as the blast wave, and about a third in the form of thermal radiation.

The Nagasaki bomb, dropped three days later, released about 92 TJ. But the explosive power of these first bombs was tiny when compared to hydrogen bombs. The most powerful US hydrogen (fusion) bomb, tested in 1954, was equivalent to 15 megatons of TNT—that is, 63 petajoules (PJ, 10^{15} joules).

This was far surpassed on October 30, 1961, when the USSR tested the RDS-220 bomb above Novaya Zemlya in the Arctic Ocean. Forty-nine years later, in August 2020, Rosatom (Russia's Atomic Energy Corporation) released a 40-minute-long film, *Top Secret: The Test of Clean Hydrogen Bomb with the Yield of 50 Megatons*. In this remarkable document, the bomb is called "the product" and is "clean" (*chistaya*), and the ubiquitous presence of antiquated analog instrumentation manned by uniformed soldiers—as well as their use of radiotelegraphy—provides unexpected archaic contrasts with the weapon's new immense destructive power. Nicknamed the Tsar Bomba, it was hung underneath a Tu-95 bomber's belly and parachuted from a height of 10.5 kilometers.

American hydrogen bomb test, 1952

The bomb exploded 4 kilometers above the ground and its explosion released 210 PJ of energy, three orders of magnitude more (about 2,300 times) than the Nagasaki bomb. The explosion created a mushroom cloud 60–65 kilometers in diameter and the flash was visible from nearly 1,000 kilometers away. And soon afterwards, the Soviet premier Nikita Khrushchev claimed that the USSR had built a bomb twice as powerful—that is, one capable of releasing more than 400 PJ of energy.

Between London in 1945—the last V-2 attack on the city was fewer than six weeks before the Nazi surrender, on March 27, 1945—and Novaya Zemlya in 1961, the maximum explosive energy of weapons rose by seven orders of magnitude (from less than 5 GJ to more than 200 PJ), matching the post-1970 orders-of-magnitude rise of component crowding on a microchip in only a third of the time (16 vs. 50 years)—and reminding us of modern civilization's terrible priorities.

The Stuff the World Is Made of

Why Green Roads Are about More than the Vehicles on Them

You are driving on a concrete road and suddenly you get on a newly laid strip of asphalt paving and instantly feel the difference. Laid in continuous stretches without any visible joints that create repetitive thudding on concrete paving, the asphalt surface is pliable, smooth, vibrations and noise are reduced, tires get better traction. A similar effect on concrete can be achieved only when its surface is treated by diamond griding. But calling that smooth paving an asphalt road is an even greater misnomer than it would be calling a concrete surface a cement road: there is much less asphalt in "asphalt" than there is cement in concrete!

Cement (made by heating limestone, clay, and other minerals in rotary kilns) makes up typically just 15 percent of concrete, the rest being water and aggregates (gravel, sand, finer particles). When you drive on an asphalt road, as much as 96 percent of the material under your wheels is the same, or similar, aggregates (providing strength and load-bearing capacity) mixed with just 4–6 percent asphalt, whose function is to bind them and confer flexibility, durability, and water resistance. "Bitumen road" or "asphalt concrete pavement" would be more accurate terms—but after generations of inaccurate labelling it would be even more confusing.

Which one is better: concrete or asphalt? Properly laid

concrete is more durable (a lifespan 2–4 times that of asphalt), more resistant to rutting, and better able to withstand repeated heavy loads, but damaged concrete is not easily patched (entire slabs must be cut out and replaced), the ride is not as smooth (both bumpier and noisier), and the initial cost of a concrete road is higher. Asphalt roads can be built faster, withstand higher temperatures without buckling, are smoother, do not last as long, and are more vulnerable to rutting by alternating freeze-thaw—but their initial cost is lower. And that is why in the US more than 70 percent of interstate highways are paved with asphalt, with concrete more common on roads with heavy traffic and in northern states, and why asphalt is a commodity in increasing demand: the global market is expected to go from about 120 million tons in 2020 to 200 million tons by 2030, with the highest gains in China, the US, and India.

Asphalt (bitumen) is the blackest, heaviest, and stickiest fraction of crude oil, the mixture of heavy hydrocarbons that remain "at the bottom of the barrel" at temperatures above 425°C after the progressively heavier products (from light gasoline and kerosene to heavy fuel oils) have been separated in refineries by distillation—leaving behind a sticky, semisolid, highly viscous substance. Every crude oil refinery is thus a potential source of asphalt, and the US refineries supply about 90% of domestic demand. There is also natural asphalt, known from antiquity, extracted in liquid or semisolid form from asphalt lakes (the most famous being La Brea in southwestern Trinidad) or in solid form from sedimentary rock formations.

Asphalt is used in roofing and waterproofing in construction, and in the production of insulation materials, adhesives and sealants, but road and pavement uses dominate. Asphalt is commonly mixed with such additives as polymers, waxes, and thermoplastic rubber to make it stronger and more workable.

Unfortunately, some asphalts have been illegally mixed with a heavy residue (called REOB, re-refined engine oil bottoms) left after recycling lubricating oils: the residue contains chemical additives (including trace metals) that cause premature surface cracking in cold temperatures. And much like concrete, asphalt can be recycled. A survey done by the National Asphalt Pavement Association shows that in the US during the early 2020s about 95 million tons of reclaimed asphalt pavement (RAP) was reused in new pavements—making up just over 20 percent of new mixtures, and conserving a similar mass of virgin aggregates and nearly 5 million tons of asphalt binder as well as freeing up landfill space.

US prices of asphalt paving mixtures have increased from about $130 in the year 2000 to $400 by 2024, and about $2.5 billion of refinery asphalt is now traded every year, with the US, Canada, Italy, and Russia being the largest exporters but the US and Canada also being the largest importers—buying nearly the same amount as they sell. Given the remaining crude oil resources (including the enormous deposits of heavy bitumen in the Canadian oil sands) there is no danger of running out of asphalt, but with some 95 percent of the global output coming from refineries, what would happen to its supply in a world running completely on renewable energies?

We would need substantial quantities of materials that are now derived from crude oil or coal even in a zero-carbon world: not only asphalt for paving, but also tens of millions of tons of lubricating oils for the smooth operation of machinery, and carbon black for tires, printing, and the electrical industry. When larger-scale crude oil refining began during the late 19th century—before automobilization led to the rise of paved roads—refiners, producing mostly kerosene and lubricants, were giving asphalt away for free. What would they do with the gasoline, kerosene, and heavier oils they would need to produce to make the still-needed asphalt if they could not even give them away?

The extraction of natural asphalt could be vastly expanded, but that would not succeed without concurrent advances in assuring a uniform product quality and the specific diversity now guaranteed through refining. For example, the Cenovus Lloydminster Refinery, Canada's largest asphalt producer in Alberta (owned by Husky Energy), now turns out 30 different grades of asphalt with guaranteed compositions and properties, something that would be hard to match when extracting a large natural asphalt deposit. In a decarbonized world there may still be a place for that black, heavy, viscous hydrocarbon mixture that makes smooth driving possible—even in electric cars.

Asphalt paving machines in action

A Grain of Truth

For most urban dwellers in the colder parts of the northern hemisphere, sand now evokes above all images of subtropical and tropical beaches where they will be flown for winter vacations by packed jetliners; and for millions of people playing golf, sand is associated with dreaded bunker shots landing in sand-filled depressions surrounding the greens. But those travelers and weekend players are mostly unaware of the many fundamental roles sand plays in enabling the functioning of modern civilization, from its structures to its clean water supply. Modern urbanites, no matter if they live in old brick houses or in new high-rises, inhabit structures that could not have been built without sand (and that includes windows: see "From Glass Beads to Massive Windows," p. 96). Without sand there would be neither the concrete nor asphalt paving that makes rapid modern road transport possible (see "Why Green Roads Are about More than the Vehicles on Them," p. 83), nor any glasses to drink from—indeed no clean water in those glasses. These are the facts.

Bricks are largely clay (50–60 percent) for plasticity and sand (20–30 percent) to provide texture and to prevent excessive shrinkage and cracking during the drying and firing in kilns (lime and iron oxide make up the rest of the raw materials). Concrete is a mixture of cement (material produced by the high-temperature firing of limestone, clay, and sand in long rotary kilns), water, and coarse (gravel) and fine (sand) aggregates; these aggregates make up 60–80 percent of the total mass, and between a quarter and a third of all concrete structures is sand. Not every sand will do, however. Grain sizes should range between 0.075 and 4.75 millimeters, and the sand grains should not be rounded but sharp-sided and angular, to ensure better interlocking with other ingredients. That is why sea sand, even when dried and with salt removed, is not a good concrete-making aggregate, because its smoother and rounder grains cannot make a strong finished product: river sand, pit sands, or manufactured sand (crushed stone) can do that. Sand also makes up a similar share (25–40 percent) of asphalt paving, which contains up to 96 percent of aggregates.

Sand is constantly produced through the natural diminution of rocks driven by weathering, erosion, and transportation by water, glaciers, and wind. Quartz-rich sand comes from the weathering of granitic and metamorphic rocks; marine (carbonate) sand comes from corals, shells, and skeletons; gypsum sand originates from evaporite basins in dry regions; and volcanic eruptions produce mostly dark and heavier basaltic or glassy sand, as seen on Hawaii's beaches. The erosion and transportation of sand by wind (including very long distances, as the Saharan sand repeatedly reaches the Caribbean), streams, waves, currents, and glaciers create sedimentary formations, dunes, and beaches. The global resources of sand are enormous, but local overexploitation has become increasingly common.

Detailed American statistics explain the typical annual disposition of sand. During the early 2020s, about 45 percent of sand was used as concrete aggregate, about 20 percent went into road base, coverings, and road stabilization, 13 percent ended up as construction fill, 12 percent was used to make asphaltic mixtures, and the remaining share (about 10 percent) has always had a wide range of industrial applications, above all in glassmaking and foundries, in filtration, in plaster mixtures, as roofing granules and railroad ballast, and in some cold regions for road ice control. Glassmaking, ceramics, and foundry sand (to make molds for metal castings) remain the largest industrial uses everywhere except in the US. By weight, the raw materials for glass are 70–75 percent silica sand (SiO_2 rich), 10–15 percent soda ash (sodium carbonate), and 10–15 percent limestone (calcium carbonate), with small amounts of additives, mostly for color.

In the US, about three-quarters of the annual demand for industrial sand now comes from hydraulic fracturing (fracking), the extraction of oil and gas by injecting a pressurized mixture of water, sand, and additives into hydrocarbon-bearing strata. The composition of fracking fluids varies according to specific extraction demands, but by weight water always dominates (80–90 percent), with sand—used as essential proppant, the material that keeps rock fractions open and enables hydrocarbon to exit through tiny openings—making up 9–15 percent of the total mass. Fracking sand properties are modified by additives, to reduce friction, lower the surface tension of water, and prevent the formation of emulsions that would block the escaping oil and gas. Sand is also indispensable for packing and cementing oil and gas wells. Other smaller but irreplaceable uses of sand include as abrasives for blasting and sanding and, commonly, for municipal water filtration.

So now you know: if you are reading this sitting in a house with concrete foundations and brick walls warmed by natural gas produced by fracking, holding a glass of cold water and looking out of a large window or at a mobile, TV, or computer screen, none of these amenities would be possible without appropriately selected sand. Inevitably, there is an

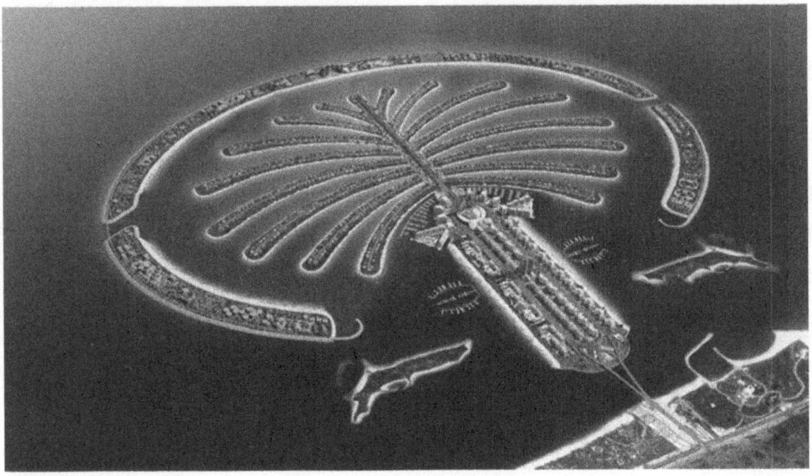

The Palm Jumeirah: sitting on sand

environmental price to pay for all these uses. Not only is the production of sand highly energy-intensive, but the vast volumes of annually removed material result in a substantial fuel and electricity demand for its excavating, dredging, transportation, processing, and its use in construction. The worst environmental impacts accompany massive, and too often excessive, sand extraction in rivers and along ocean coasts—as they affect water flows, increase erosion, and destroy wetlands.

Among the world's most prominent land reclamation projects, Dubai's Palm Jumeirah (an area of 5.6 square kilometers) required 5.5 million cubic meters of rock from quarries and 94 million cubic meters of sand excavated from the deep seabed. Construction of Hong Kong's Chek Lap Kok airport reclaimed 9.38 square kilometers of land from the sea by displacing about 59 million cubic meters of sand, while Singapore's Changi airport was built on more than 40 million cubic meters of reclaimed sea sand. Such projects would be impossible without deploying large, specialized ships (trailing suction hopper dredgers). And millions of tourists traveling to Honolulu's Waikiki Beach are unaware that a century ago it had to be replenished with sand imported from California's Manhattan Beach, while the most recent sand replenishment of that famous beach removed sand from the seabed between 500–1,000 meters offshore and pumped it back onto the beach as water-and-sand slurry. You might recall some of these remarkable facts next time you walk on a beach—or just look through a window, or land on a runway.

You Might Not Have Thought Much about Silicon—but It Is Indispensable

By mass, silicon is the second-most important element in the Earth's crust (oxygen bound in many minerals is the first), but in nature it does not exist in a pure (elementary) state. Silicon makes up 47 percent of silicon dioxide (SiO_2, silica), the dominant compound in most non-tropical sands (see "A Grain of Truth," p. 87). Silica remains one of the three essential ingredients (together with sodium and calcium carbonates) in glassmaking, first practiced in ancient Egypt and now one of the indispensable modern industries. But other industrial uses require elemental silicon that must first be isolated from silica. The oldest—and in mass terms the largest—of these uses is the metallurgical-grade silicon produced through the deoxidization of silica with carbon. Crushed silica is mixed with coal, coke, or charcoal and loaded into arc furnaces, where electricity passing between graphite electrodes creates temperatures of up to 2,000°C, to produce molten silicon that is cast to make solid blocks for further uses.

Metallurgical-grade silicon is 99 percent pure, and more than 8 million tons of it are consumed annually, mainly in producing ferrosilicon (to deoxidize steel) and aluminum alloys (to enhance their strength), and to make silicones—synthetic polymers used since the 1940s as lubricants, resins, insulators, and water repellents. For common use, silicones come either in small plastic tubes or in larger rigid plastic cylinders loaded into a caulking gun, handy whenever they are needed to seal sinks, tubs, showers, fixtures, or countertops. Besides that, silicones have an enormous range of commercial uses, from antiperspirants, deodorants, shampoos, and lotions (they impart that smooth feel) to computer keypads and keyboards, and from non-stick bakeware and cookware to sealing joints in buildings and bridges as well as the doors, windows, fuel tanks, and engine gaskets in jetliners. But even 99 percent purity is totally unacceptable for the other two major classes of silicon use: solar (photovoltaic) panels and electronics.

That is because the pure element is a relatively inexpensive semiconductor, a material whose properties are between those of electricity conductor (such as copper, gold, aluminum, or iron) and insulator (glass, plastics, wood). After adding small amounts of impurities or applying a specific voltage, silicon can act either as an insulator or a conductor—providing the most suitable substrate for producing the microprocessors (microchips) that are indispensable for the functioning of modern civilization. Solar-grade silicon has a purity of 99.9999–99.999999 percent; electronic-grade silicon is up to 99.999999999 percent pure. An elaborate process is needed to purify metallurgical-grade silicon to such degrees, turning it first into trichlorosilane (by a process invented by Siemens during the 1960s), then decomposing it to yield the element, and then adding tiny amounts of impurities to modify its conductivity.

To make thin silicon wafers, this polycrystalline silicon must first be converted into large crystals. The way to do that was discovered accidentally in 1916, more than three decades before the beginning of the electronic silicon era, by Jan Czochralski, a Polish metallurgist working in Germany. As he dipped his pen into a small container of molten tin, he lifted a thin thread of metal, and by experimenting he discovered that—when combined with a slow rotation and gradual decrease of temperature—this lifting could make metallic crystals of tin, lead, and zinc that were up to 1.5 centimeters long and a few millimeters in diameter.

In 1948, when germanium was to be used in the first semiconductors, Gordon K. Teal and J. B. Little at Bell Labs turned to Czochralski's method to grow its pure crystals, and two years later Teal and Ernest Buehler produced the first pure but tiny silicon crystals. By the beginning of the 21st century the maximum crystal diameter had reached 30 centimeters, and while larger crystals (45 centimeters across) have been developed, their production is not yet commercialized. With their length ranging between 1 and 2 meters, the 30-centimeter cylindrical jet black super-pure silicon ingots weigh between 165 and 330 kilograms.

Their cutting starts with trimming the irregular ends followed by cutting with a thin (120–150 micron) wire coated with diamond particles. This produces slices of 150–300 microns that are ground to exact thickness, and have their edges rounded and surfaces smoothed by etching and polishing to be ready for producing integrated circuits. That involves an elaborate sequence—polishing, cleaning, the application of light-sensitive chemicals, shining ultraviolet light through a mask to form circuit patterns, etching the exposed layers, doping (the implantation of minute impurities), vapor deposition for coating and polishing—and these steps are repeated to produce multiple layers of transistors, capacitors, and wiring on the wafer.

Just about everything in our world—from the laptop used to write this book to the kitchen oven I use for baking, from intercontinental flights to driving a car, from diagnostic

Silicon crystals and wafers

imaging to designing bridges—depends on microprocessors (electronic design, control, and management are now truly ubiquitous), and none of this would have been possible without discovering the ways to isolate silicon from silica, to purify it to 10 nines, to grow its massive crystals, to slice them with diamond-coated wires, and to use the wafer as substrates for making microchips in the world's most expensive factories. And much like everything else in the modern world, this has become a matter of mass-scale production.

About 150,000 tons of electronic-grade silicon (upgraded from the metallurgical silicon) are used to make more than 8.5 million square meters of silicon wafers (roughly an area of 3 x 3 kilometers). Those wafers, turned into more than a trillion microchips, now control everything from toilets, dolls, and pacemakers to national electricity grids and global weather monitoring. And we are in no danger of running out of the raw material that makes the best semiconductor: a few exceptions aside, the Saharan dunes are composed of pure quartz sand and nearly half of it is silicon.

From Glass Beads to Massive Windows

Plate glass is another taken-for-granted ubiquitous material of the modern world. Much like concrete or steel, it is all around us, and much like with cement (the indispensable constituent of concrete) and steel, most people would not be able to explain how this transparent material is made. All these man-made materials go far back: the first steel weapons date to before 1000 BCE, some Roman concrete structures still stand (most notably, Rome's Pantheon), and the oldest Egyptian glass beads were made about 4,000 years ago. But while open-hearth furnaces transformed artisanal steelmaking into a large-scale commercial activity by the 1870s, and while large-volume rotating cement kilns were introduced during the 1890s, turning out large pieces of plate glass became possible only after the Second World War thanks to the successful application of an older idea of floating molten glass on a large bath of molten metal (lead or tin).

The first larger pieces of flat glass (broad sheets) to be used for windows were made in England before the end of the 13th century, by blowing molten glass into elongated tubes and then by splitting and flattening the resulting cylinders. Later came so-called crown glass, blown into a hollow sphere and then reheated and spun into a disk with a diameter of up to 1.8 meters. These artisanal methods produced limited sizes of unevenly thick plate glass, and the first mechanical process came only at the beginning of the 20th century, with cylinders drawn from a molten glass tank, cooled, cut, reheated, and flattened. But it was only in 1952 that Alastair Pilkington, a British engineer and businessman, invented an efficient and fully mechanized process to make directly large flat pieces of glass of a desired thickness.

After seven years of working with Kenneth Bickerstaff to develop a mature, patentable commercial version, it was made public on January 20, 1959, and before the end of the year they had got the US patent for "the manufacture of flat glass in continuous ribbon form." The process begins with mixing the ingredients for soda-lime glass: besides the ubiquitous sand (SiO_2, making about 73 percent of the charge), the mix is about 13 percent sodium oxide (Na_2O), 9 percent lime (CaO), and 4 percent magnesium. The combustion of natural gas (or of a liquid fuel) melts these solids in a glass-making furnace, and the molten glass (at 1,050°C) is then poured from a ceramic spout, passed between sizing rolls to form a ribbon of the desired thickness, and then floated over a six-centimeter-thick layer of molten tin (at 230°C).

The molten tin baths are 4–8 meters wide with lengths of up to 60 meters, and the tin's oxidation is prevented by keeping the metal under a protective nitrogen (90 percent) and hydrogen atmosphere. As the molten glass spreads, it achieves uniform thickness, and the process can be regulated to produce sheets as thin as 0.12 and as thick as 25 millimeters. After the formed sheet cools to about 600°C, its cracking is prevented by passage through a long furnace ("lehr" is its technical name) where it is gradually cooled at a precisely controlled rate. Once cooled, the glass can be cut into desired shapes (mostly rectangles) for distribution.

This new method quickly conquered the global plate glass market as all the world's leading makers licensed the Pilkington process during the 1960s. Most plate glass producers offer sizes of up to 3.3 x 6 meters, but in 2021 Tianjin NorthGlass supplied the world's largest windows, for the lobby of the Taikang tower in Beijing, measuring 3.1 x 17.1 meters with a total area of 50.87 square meters.

Much like large modern blast furnaces, which keep producing molten iron for more than a decade before they are extinguished and their interior coating is replaced to start a new iron-smelting campaign, float glass plants work nonstop for 11–15 years, producing (depending on the size of the tin baths and number of lines) anywhere between 50 and more than 1,200 tons of glass per day. Most plants making float glass have a single line, but the largest float glass factories (now all in China) are operating more than 10 lines at one site. Many countries, including the Netherlands, Portugal, and Romania, have a single line that serves mostly the local plate glass market.

The US now has 23 plants with 31 lines (mostly in Ohio, Illinois, Tennessee, and Texas), while China, by far the world's largest glass producer, has more than 50 factories and 190 lines. This means that even when adjusted for its much larger population size, China now tops such formerly dominant plate-glass-making powers as Germany, France, and Italy, and only Poland, the Czech Republic, and Luxembourg are Europe's relatively more "glassy" economies in per capita terms. In contrast, several major nations have no plate-glass capabilities; they include not only Africa's three most populous states but also Canada (where the last plate glass plant, operating in Quebec, was shut down in 2005) and Sweden.

Making plate glass is not only fairly energy-intensive, but today's standard practice simply cannot do without fossil fuels.

Making molten float glass takes almost as much energy (mostly natural gas) as does making steel. It is thus yet another industrial process (along with producing primary iron, ammonia, cement, and plastics) that cannot be easily decarbonized—that is, converted from natural gas to solar or wind (or nuclear) electricity.

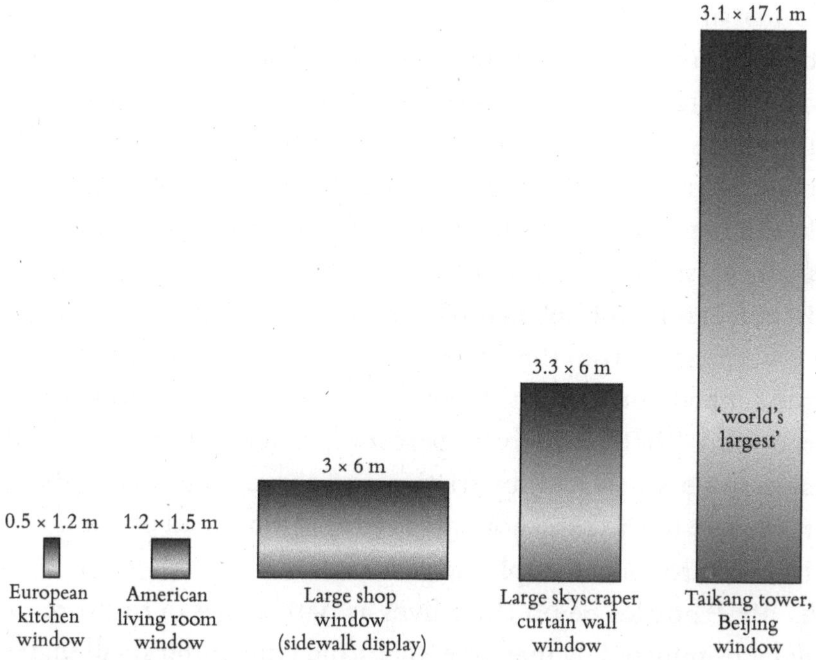

Sizes of window

Endless Streams

Unfortunately, discarded polyethylene terephthalate (PET) bottles are among the most visible eyesores and pollutants of modern civilization, produced and thrown away by the hundreds of billions every year—spilling from garbage cans on sidewalks of megacities, lying discarded on beaches and in ditches, forests, and fields, floating on tides, aggregating in giant plastic masses caught in the shallow waters of tropical islands, and bobbing against the shores of beaches thousands of kilometers from the closest large settlements. When empty, the capped bottles float; when filled with water or broken up, they sink (PET is nearly 40 percent heavier than water)—and as they get worn away by friction and abrasion they contribute massively to the enormous (and rising) global load of countless forever-around microplastic particles.

PET bottles begin their lives as natural gas, a mixture of gases dominated by methane (CH_4) and containing small shares of ethane (C_2H_6) and propane (C_3H_8). The two heavier gases are separated from methane, mixed with steam, and heated in furnaces at 750–950°C to produce ethylene (C_2H_4). Ethylene is then oxidized to make ethylene oxide, which is reacted with water to make ethylene glycol. The other key ingredient to synthesize PET is terephthalic acid (TPA), made by catalytically (in the presence of cobalt or manganese) oxidizing para-xylene derived from natural gas and then purifying and drying the product. Ethylene glycol and TPA are reacted to produce short-chained monoethylene glycol terephthalate which is heated in a vacuum to form a long-chain PET polymer.

PET was commercially introduced during the 1940s, and for decades it was best known as the common synthetic fabric polyester, or under such brand names as Terylene or Dacron—paragons for making cheap shiny suits. Other PET uses have included flexible packaging (above all, the ubiquitous and maddeningly hard-to-open blister packs), material for 3-D printing, containers for high-pressure gas cylinders, waterproofing for undersea cables, coating for ropes to reduce their wear, fabricating durable parts for automotive and aeronautical uses, and as backing for pressure-sensitive adhesive tapes. When combined with glass fiber, PET makes thermoplastic resins found in electrical appliances, and one application now covers increasing areas of the Earth's surface: a stretched form of PET forms the back sheets of photovoltaic modules.

The global demand for PET was about 33 million tons in 2025 and it is expected to rise to 42 million tons by 2030, with a third of the total going to polyester fibers for clothes, upholstery, and industrial uses, and 40 percent to the production of beverage bottles as well as containers for other liquids. This application began in the early 1970s and has now conquered the global market for both still and carbonated water, as well as for many brands of sugary soft drinks. Using the material is a good physical fit because PET is clear, strong, flexible, and light, having only half of the density of bottle glass (1.38 g/cm^3 compared to 2.6 g/cm^3). Like all plastics, it is recyclable: it carries the digit 1 (♳) of the resin identification code, and the good news is that it has been recycled more assiduously than any of the other six categories of plastics. But that is good news only in relative terms, because the recycling rates of other plastics are depressingly low: PET recycling rates look especially good when compared to the billions of polyethylene plastic bags that are overwhelmingly discarded after a single brief use.

About 600 billion PET bottles are now produced worldwide every year, and that means that about 1.1 million bottles are bought every minute and nearly 20,000 bottles are discarded every second! As with other containers, the average weight of PET bottles has decreased over time, and for a 1-liter bottle it now ranges from 30 to 50 grams for still beverages and 50–60 grams for carbonated drinks; for half-liter bottles the mean is about 25 grams. Even when assuming the lowest unit weight (for still-water bottles) this would imply discarding close to a ton of PET bottles every second! Recycling has been improving but it has a long way to go. In the US, PET bottle recycling has been only around 30 percent of annual production (75 percent in Maine, 3 percent in Alaska). This means that in the US every year some 35 million single-use PET bottles are landfilled, discarded in the open, or incinerated.

To reach the US 2030 target of having 50 percent recycled PET in new beverage bottles, this would require tripling the weight collected in 2022—an unlikely prospect. In Europe, recycling rates of PET bottles have been much higher, averaging about 75 percent for EU member states, but in the rest of the world the recycling rates are below 15 percent. More fundamentally, bottle PET can be recycled only two or three times to make new bottles, and as the polymer chain breaks down the old PET must be "downcycled" (for such non-food applications as carpets or clothes) or mixed with new PET to assure structural integrity. Only expensive (energy-intensive) decomposition to its base monomers could make indefinite repolymerization possible.

The cost of bottled water is ridiculous: while tap water

costs a fraction of a cent per liter, tap water in a PET bottle goes typically for a couple dollars. Much worse, the environmental impact of discarded PET bottles is depressing. Waste carried by the Mississippi is now 75 percent plastic, and PET bottles have been the second-most common waste item encountered when surveying the river's delta. Plastic bottle caps are the second-most common waste item polluting US beaches (cigarette butts lead); PET bottles rank fourth. The time for full PET bottle decomposition is nearly 500 years, leaving behind microplastic particles that are already an enormous source of uncontrollable pollution, with some 75 trillion pieces in the oceans and on the beaches. All marine organisms (from crustaceans and turtles to fish and seabirds) now have microplastics in their bodies—as do we.

Plastic waste in the Maldives

Why E-Waste Is Worth More than Its Weight in Gold Ore

In the mid-2020s, the average mass of a mobile phone was 180 grams (a range of 120–250 grams), with the impact-resistant display screen made of Corning's Gorilla Glass being the heaviest component (50 grams), followed by various plastics (20–30 grams), aluminum (25 grams, mostly in the casing), steel (20 grams, mostly in the frame), and copper (17 grams). These are small aggregates, but given the annual global sales of mobile phones they translate to large totals. These sales surpassed 100 million units in 1997, 1 billion in 2009, the peak in 2018 was 1.556 billion, and 1.4 billion units were sold worldwide in 2025. That total now incorporates some 70,000 tons of special glass, 35,000 tons each of aluminum and plastics, 28,000 tons of steel, and about 25,000 tons of copper annually.

Also some 40 tons of gold (about 0.025 grams per phone), more than 1,000 tons of toxic heavy metals (above all, lead in the solder, mercury and cadmium in batteries, and chromium in screens), and more than 200 tons of rare earths (primarily in magnets and displays). And to focus on the brain of the device, to make every kilogram of electronic circuit boards (assemblies accounting for about a fifth of a phone's mass) requires 250–450 grams of copper, 15–60 grams of nickel, 1.5–4 grams of silver, and about 1.5 grams of gold. Obviously, in a rational world these devices would be designed to last long and to remain functional by providing suitable automatic updates: five years should be no problem, a decade could be possible.

And at the end of their useful life, they should be assiduously recycled: after all, besides some 25 kilograms of gold, 1 million discarded mobile phones contain 250 kilograms of silver and more than 17,000 kilograms of copper. Instead, amidst all the headlines about new economies being green and "sustainable" (I have been allergic to this adjective for decades: sustainable across what time span—a decade or a century?—and across how many usage categories?), most mobiles not only end their lives prematurely but are not recycled—and everybody is guilty.

Phone companies incessantly push new models through relentless advertisement, special offers and discounts, and they do not support aging models by providing ongoing operation and security updates. Too many consumers, driven by the me-too attitude to own and to boast, choose to dump their perfectly good devices for the latest model. And governments have not enacted laws that would require the provision of extended operation and mandatory recycling. As new devices keep coming along, perfectly good older units are prematurely discarded, on average, after just two and half years of use. As a result, the global total of dumped phones is now approaching 6 billion units a year, or more than 1 million tons of electronic waste.

And, unfortunately, that is only a smaller part of the total mass of e-waste generated globally. If I were to discard my mobile phone, I would add 190 grams to the e-waste stream but its two accessories (a charger at 85 grams and a portable power bank at 220 grams) weigh more than the device. My iPad weighs nearly 500 grams, its charger adds 85 grams, the laptop used to write this book has a mass of 1,850 grams, its cord (with a transformer) adds 300 grams—altogether more than 3 kilograms of personal electronics. To this must be added several categories of even more massive e-waste ranging from audio and video equipment to TVs, printers, copiers, and microwave ovens, as well as the now-ubiquitous light-emitting diodes and batteries, and the latest (and fastest-growing) addition includes many kinds of photovoltaic panels.

This e-waste contains higher concentrations of costly, precious, and toxic elements than you could find anywhere in nature. The same amount of gold that could be found in 12 tons of gold-bearing ore can be extracted from 1 ton of e-waste. Chilean copper ore is just 0.6 percent copper, so it is necessary to extract 165 tons of ore to get a ton of the metal—but many kinds of e-waste are 5–10 percent copper, and printed circuit boards are as much as 14 percent copper, more than 20 times as copper-rich as Chilean ore. And discarded printed circuit boards contain as much as 1,500 grams of gold per ton, 40 times the metal's concentration in ores mined in the US. This makes recovery of metals from e-waste pretty convincing.

The global estimates are, necessarily, only approximate, but recently e-waste generation has surpassed 60 million tons a year—of which less than 25 percent was formally collected and recycled. The detailed partitioning of this waste stream is unknown, but it has been widely estimated that up to 40 percent of old mobile phones are not immediately discarded but kept at home, part of a wider phenomenon of modern hoarding. In 2024 the Global E-waste Monitor concluded that high-income countries average 109 items of electrical (including refrigerators, stoves, and washing machines) and electronic equipment (excluding lamps) per capita, while in Europe an average household has more than 70 electronic items of which

11 are no longer in use or are broken, and every year several kilograms of unused electrical and electronic products are hoarded before they are eventually discarded.

Less than 25 percent of the world's electrical and electronic waste is now recycled, with continental rates ranging from more than 40 percent in Europe to less than 1 percent in Africa, with most laws relying on extended producer responsibility (making producers accountable for the entire life cycle of their products). Common metals are the most recycled materials, and recycling is also relatively common for the valuable platinum group of metals. China is by far the largest generator of electrical and electronic waste with about 13 million tons a year; the US leads in per capita terms (about 22 kilograms a year). Significant amounts of e-waste are exported, and not only from richer to poorer countries (for illegal dumping) but also the other way around—to France, Germany, Sweden, Japan, or South Korea, countries with advanced capabilities to process the valuable materials safely and profitably. But it all amounts to too little, as we need (at least) to quadruple the global recycling rate to get e-waste under control. Given the consequences, compulsory (deposit-driven) recycling might be the only effective solution.

Moving Things and People

Bigger and Bigger Cars

In 1920, US car sales entered the period of their fastest expansion. In that year Americans bought 1.9 million new vehicles; five years later the total rose to 3.75 million, with the three versions of Ford's Model T (Runabout, Coupe, Sedan) being the bestsellers. Their mass was, respectively, 630, 717, and 794 kilograms, powered by engines rated at 20–22 horsepower (15–16 kilowatts). And then the cars became more powerful, the design shifted to fully enclosed (and soon to streamlined) metal cabins, and many models were acquired. Just before the Second World War, Ford's 1941 Deluxe weighed almost 1.5 tons (1,480 kilograms), and after the war came decades of ever larger, longer, and more powerful models adorned with outrageous tailfins, bulging lights, and massive chrome grilles. In 1958 a Cadillac, the decade's signature car, weighed more than 2 tons (2,030 kilograms) and had a power of 231 kilowatts.

In contrast, post–Second World War European and Japanese cars were tiny and light. The popular, affordable archetypal French Citroën 2CV weighed just 510 kilograms, Fiat's Topolino (meaning "a small mouse") 550 kilograms, and Volkswagen's Beetle as little as 730 kilograms; and in 1973 when Honda began to sell its Civic in North America it weighed just 697 kilograms. But in the US these cars were just curiosities: there was no incentive for downsizing when the country remained the world's largest producer of crude oil, with gasoline selling for 35 cents per gallon (the average price in 1969).

Things changed suddenly in the wake of OPEC's first round of oil price increases in 1973. When the US Environmental Protection Agency began monitoring the mass of new car models in 1975, a typical European or Japanese car was just 750–900 kilograms but the mean weight of US sedans and light trucks (1,840 kilograms) was approaching 2 tons: between 1920 and 1975 the typical American car became more than 2.5 times heavier. The subsequent (1979–80) near-quadrupling of the world oil price brought a prompt return to lighter vehicles, and by 1981 the average mass of new US cars had declined to 1,450 kilograms. At that time nobody predicted how short-lived this retreat would be. By 1985, the world oil price returned to pre-1979 levels, and this resulted not only in the return to heavier car models but in the introduction of an incongruously named class of sports utility vehicles (SUVs).

Subsequent oil price fluctuations had no effect on the SUV's North American—and soon also worldwide—popularity. As the share of sedans kept on declining and as the shares of SUVs and pickup trucks kept on increasing, the inevitable had to happen. By 2004 the average mass of new US passenger vehicles had surpassed the 1975 level to reach a new record of 1,860 kilograms—and since that time the three variables of car size (vehicle mass, footprint, and engine power) have kept on rising, slowly but relentlessly. By 2022 the average mass was very close to 2 tons (1,964 kilograms), vehicle footprint rose from 4.5 to 4.8 square meters, and installed power increased from 156 to 218 kilowatts.

And America's bestselling vehicle (ever since 1976!) has become considerably heavier and more powerful: the 2025 model of the Ford F-150 pickup truck weighs as much as 2,659 kilograms and has 320 kilowatts of power. Between 1920 and 2025 the average mass of new US vehicles had nearly tripled, and the average power rating rose nearly 15 times. The average mass of European cars is still lower, but it too has increased, even for compacts—from about 800 kilograms by 1970 to just over 1,400 kilograms by 2020. This runs, obviously, counter to the common historic trend of making products less heavy by introducing better designs, by using lighter (yet stronger) materials, and by miniaturizing components or entire assemblies.

Examples of this trend abound, from packaging to sports equipment, from TVs to phones, from gas furnaces to airplanes: aluminum cans (redesigned from an initial 80 grams to just 13 grams), tennis rackets (wood 400 grams, graphite 250 grams), TVs (cathode-ray tube 50 kilograms, LED 15 kilograms), mobile phones (from 1,000 grams for the earliest models to 150 grams), household natural gas furnaces (from

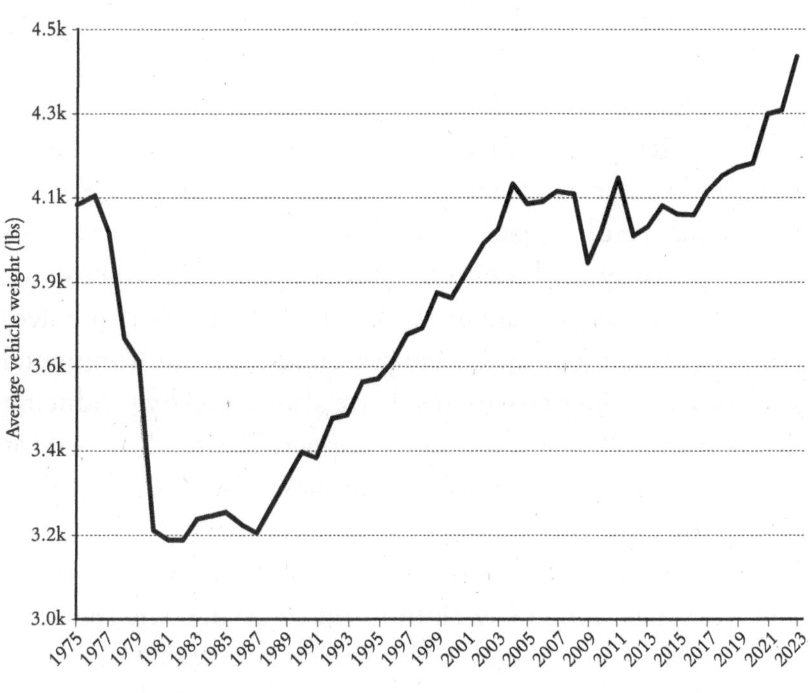

Average weight (lbs) of US cars since 1975

100 kilograms for cast iron and heavy steel to 50 kilograms for high-strength alloys), airplanes (from aluminum and steel to carbon fiber and titanium—about a 20 percent mass reduction).

And electric vehicles will not turn the trend—just the opposite. Of course, they do not emit any combustion gases, but because of their massive battery packs (weighing 400–500 kilograms) they are inherently heavier. For midsize cars the differences are between 450 and 540 kilograms, for SUVs 450–600 kilograms—that is, as much as the total mass of post–Second World War European cars! And the mantra about heavier vehicles being safer needs a more nuanced perspective. In general, they are safer in single-vehicle crashes (they are better at absorbing and distributing crash forces), but when two heavier cars collide their greater combined mass will result in a higher total kinetic energy that will have to be absorbed by their bodies; and, above all, heavier vehicles pose increased collision risks to all lighter vehicles and to all pedestrians.

And heavier electric cars have another undesirable weight effect: additional pressures on the road (hence more friction wear) and the near-instantaneous torque delivered by electric motors that results in 20–50 percent higher tire wear, burdening the roadways and their vicinities with far more microplastic dust (including the smallest particles that are easily inhaled) than exhaust emissions. Of course, significant environmental burdens and higher energy needs are also caused by producing more materials (steel, aluminum, copper, plastics, lithium) for heavier cars and more gasoline and diesel fuel (or more electricity) for powering them.

Yet there is no insurmountable reason why we could not have very safe vehicles weighing up to 40 percent less than today's American average: we have the theoretical understanding, engineering capabilities, and superior materials to do so.

Moreover, these vehicles, properly maintained, could operate for 20–25 years, much like our jetliners do, bringing additional environmental benefits. There is nothing smart or desirable about ever-heavier vehicles and frequent model changes. Will we ever learn?

Ford's F-150 pickup truck

An Underappreciated Technical Revolution

Because of its effects on weight, volume, and efficiency in movement, size matters much more in transportation than in stationary structures and machines. Obviously, more energy is needed to move large, heavier, as well as more voluminous objects, because of their greater inertia and air or water resistance (drag and friction): all these variables increase with size and speed. Trucks, ships, and airplanes have limited volume and restricted dimensions, which put constraints on the size and shape of their cargo. Moreover, moving larger or heavier items is limited by the lack of infrastructural compatibility—that is, by the maximum capacity of roads and bridges, the height of tunnels, and the size of loading and offloading cranes. These are convincing arguments for transportation modularity: while it imposes size limits, it lowers the costs of transportation as it optimizes useful cargo space and simplifies the logistics.

Railway transport moved in that direction with the adoption of standardized cars during the latter half of the 19th century: first by introducing boxcars, and later flat, hopper, and tank cars, to carry all kinds of solid and liquid materials. The largest truck size was dictated by the width and maximum load capacity of roads and tires. In shipping the need to move crude oil and products refined from it led to the construction of larger oil tankers as well as the more capacious bulk carriers transporting such loose cargoes as coal and ores in large compartments. In contrast, even after the Second World War nearly all industrial manufactures (from apparel to tools), as well as tropical fruits, tea, and coffee, were carried in the breakbulk fashion—as goods put on board ships in individually counted units and stowed in holds as sacks, bags, boxes, bundles, or crates whose piles clogged docksides and required plenty of stevedore manpower and long hours to load and unload.

A radical departure in marine transport began only in 1956, after Malcolm McLean, an American trucker from North Carolina, began to realize his pre–Second World War dream of containerized transportation. Why not have a crane lifting the container his truck brought to the port, to load it onto the ship, to be unloaded onto another truck and moved to its destination? McLean repurposed an old wartime T-2 tanker by building steel skeleton decks that would fit metal containers, and 58 of them were moved in April 1956 from Newark to Houston. The first true container ships with 226 compartments followed in October 1957, and in 1958 Matson Navigation used containers to supply Hawaiian ports from California. The transatlantic service to Rotterdam began in April 1966, and the transpacific to Japan in 1968.

By that time the industry had assumed the now-so-familiar features: larger ships carrying greater numbers of standardized intermodal (able to fit railcars and trucks) containers—rectangular steel prisms with reinforced edges and corners and twist-lock fasteners, that could be stacked in ship holds and high above the decks, with large gantry cranes inserting hooks into their corner holes as they rapidly loaded or unloaded the vessels. Their most common size—20 feet long, 8.5–9.5 feet high and 8 feet wide—led to the TEU (twenty-foot equivalent

Ideal-X, the first container ship, in 1956

unit), which is used to count larger (40 and 45 feet) and smaller steel boxes. The introduction of refrigerated (chilled or frozen) units created opportunities for shipping many kinds of food—notably out-of-season produce—as well as pharmaceuticals and vaccines.

The global progress of the industry that was just 70 years old in 2026 has been astonishing. Ship capacities increased from fewer than 300 TEU during the late 1950s to about 1,500 TEU a decade later. The first fully cellular ship (with container compartments below deck) was launched in 1968; in 1996 Danish *Regina Maersk* was the first ship to top 6,000 TEU; since 2013, ultra-large vessels carry 18,000–21,000 TEUs; and the largest ships now carry almost 25,000 containers, more than 400 times the amount that fit on McLean's first converted tanker. These ships are powered by the world's largest diesel engines (more than 13 meters tall), travel at less than their top design speed (in order to minimize emissions), and because of their size (400 meters long, a beam of 61 meters, with 13 loading levels below and 12 levels above the deck) they cannot go through the Panama Canal and so peddle mostly between Asia and Europe and Asia and North America.

Unless you live in a port city or have traveled to one, you may not have seen any of these mega-ships, but steel containers are ubiquitously unavoidable. There are now some 40 million containers in circulation, and the total number of their moves (loading and unloading from ships and from storage, or "transshipping") is more than 900 million a year, with China having seven out of the world's top 10 busiest ports. On land, containers are carried by trains (waiting at a level crossing you may see a train consisting of more than 150 cars double-stacked with TEUs) and by fleets of trucks between and within cities. They make small mountains at railway yards, they wait to be unloaded at supermarkets and shopping centers, they lie abandoned near buildings where companies went out of business, and while single units can make a small shelter, multiple units have been converted into larger houses.

And tens of thousands of them have gone overboard when high seas dislodged their tall stacks. Many lost containers lie intact on the ocean floor, slowly rusting, some gradually disgorging their (sometimes toxic) contents, others broken up and spilling toys or appliances, some found bobbing on the surface far away from the point of loss. In the worst-ever accident, on June 17, 2013, *MOL Comfort* broke up and sank off Yemen, losing 4,293 containers. On November 30, 2020, *ONE Apus*, on its way from China to California, lost 1,816 containers, including 64 with dangerous goods. Almost everything you wear, every

The container port at Shanghai

item of out-of-season produce you eat, and every gadget you work with—from socks to mangoes to computer mice—has come to you in containers. They improved logistics, lowered the cost of products, but in aggregate they exact a considerable energy price to move among continents, and they leave a widespread environmental impact. But one thing is certain: the modern world would find it exceedingly difficult to function without them!

Flying Better, Not Faster

The first machines heavier than air flew slowly. On December 17, 1903, when Wilbur and Orville Wright made their first low-altitude hops with their canvas-and-wood plane powered by a small engine made by their mechanic, that day's longest "flight" above the dunes at Kitty Hawk in North Carolina covered 260 meters in 59 seconds—that is, 4.4 m/s (or less than 16 km/h)—before it crash-landed. In 1911 Fred Wiseman's airplane carried the first airmail from Petaluma to Santa Rosa in California at speeds of no more than 110 km/h. The first national air companies set up after the First World War used open-cockpit planes and carried a few passengers at speeds around 200 km/h. During the 1930s, the best monocoque (single-shelled) all-metal designs—best exemplified by the long-serving Douglas DC-3—and more powerful reciprocating engines raised maximum cruising speeds to 300 km/h.

After 1945, the fastest commercial propeller plane (the four-engine Lockheed L-1049 Super Constellation) reached a cruising speed of 550 km/h. The ill-fated British Comet, the first commercial jetliner (it entered service in May 1952), was powered by turbojet engines and cruised at about 725 km/h before a series of fatal accidents ended its brief run. In 1958, by the time a redesigned Comet was ready to fly, the American Boeing 707 became the first successful commercial design and it led to a still-evolving series of jetliners including the 727, 737, 747 (Jumbo), 757, 767, 777 and 787 (Dreamliner). The first versions of the 707, in service since October 1958, had a normal cruising speed of 885 km/h. In 2011, the first 787 had a nearly identical cruising speed of 903 km/h.

Why was there no speed increase in commercial air travel during more than half a century—the era marked by ever-faster speeds, be it of rail travel or machining metals, of computing and wireless communication or the delivery of goods? I hasten to add that the supersonic British-French Concorde, which flew between 1976 and 2003 and whose cruising speed was about 2,150 km/h, was a heavily subsidized bi-national vanity project that ultimately failed, rather than (as history has shown) the beginning of faster commercial flight: mass-scale, affordable commercial aviation has remained restricted to speeds just below the speed of sound (Mach 1, or 1,225 km/h at sea level).

The physics of flight explains why. As airplane speed approaches the speed of sound, the drag (the aerodynamic force that opposes an aircraft's motion) it generates rises disproportionately, leaving only a narrow speed band between making more profit by flying faster and less profit by flying too fast: a speed of about Mach 0.85 is thus both the maximum and the optimum for commercial jetliners. We do not fly any faster than a lifetime ago, when the first Boeing jet entered commercial service, but we fly better: planes are safer, more reliable, quieter, their operation is far more efficient, and (although Ryanair-style travel does not seem to suggest this) the latest planes are also more comfortable.

Reliability can be illustrated by the frequency of flight and initial service checkups, and even better by the maximum time jetliners are allowed to fly with only a single engine. After an intercontinental flight lands, a large wide-body (two-aisle) jetliner can be turned around (offloaded, cleaned, refueled, reloaded) in under two hours, and it can make as many as

60 transatlantic flights a month. Its first checkup comes only after 600–1,000 hours of flying (the equivalent of at least 80 New York–London flights), the second after 6–8 months (after more than 300 flights), and the third after fewer than two years.

Extended-range Twin-engine Operations Performance Standards (ETOPS)—defined as the maximum time an airplane can fly with a single engine over a route that contains a point further than one hour's flying time from the nearest airport—was fewer than 60 minutes for propeller planes and early jetliners. In 1976 it was extended to 90 minutes, in 1985 to two hours, and now the engines powering the Airbus A350 and Boeing 787 are so reliable that these planes are allowed to fly up to 4 hours and 10 minutes with a single engine. As a result, twin-engine jetliners can cross any ocean to any inhabited destination and do so with unprecedented efficiency: compared to the early 1960s, the latest designs consume 60–70 percent less energy per passenger-kilometer, a key gain enabling the unprecedented rise in global air traffic. These more efficient planes can commonly carry enough fuel to stay aloft for more than 15 hours, and the world's longest flight, from New York's JFK to Changi airport in Singapore, is 18 hours and 40 minutes.

As for the noise, early jetliners were powered by turbojet engines that passed all the compressed air mixed with injected fuel through the turbine to exit the exhaust nozzle at high speed: this was both inefficient and noisy. Modern jetliners have turbofan engines with a large fan compressing the incoming air but directing most of it around the engine. This bypass air exits the nozzle at about half the speed of the air that goes through the engine's combustor, resulting in higher efficiency and less noise. And the cabins? In older jetliners the air was pressurized to the equivalent of about 2,100 meters above sea level, while the latest models provide higher pressurization

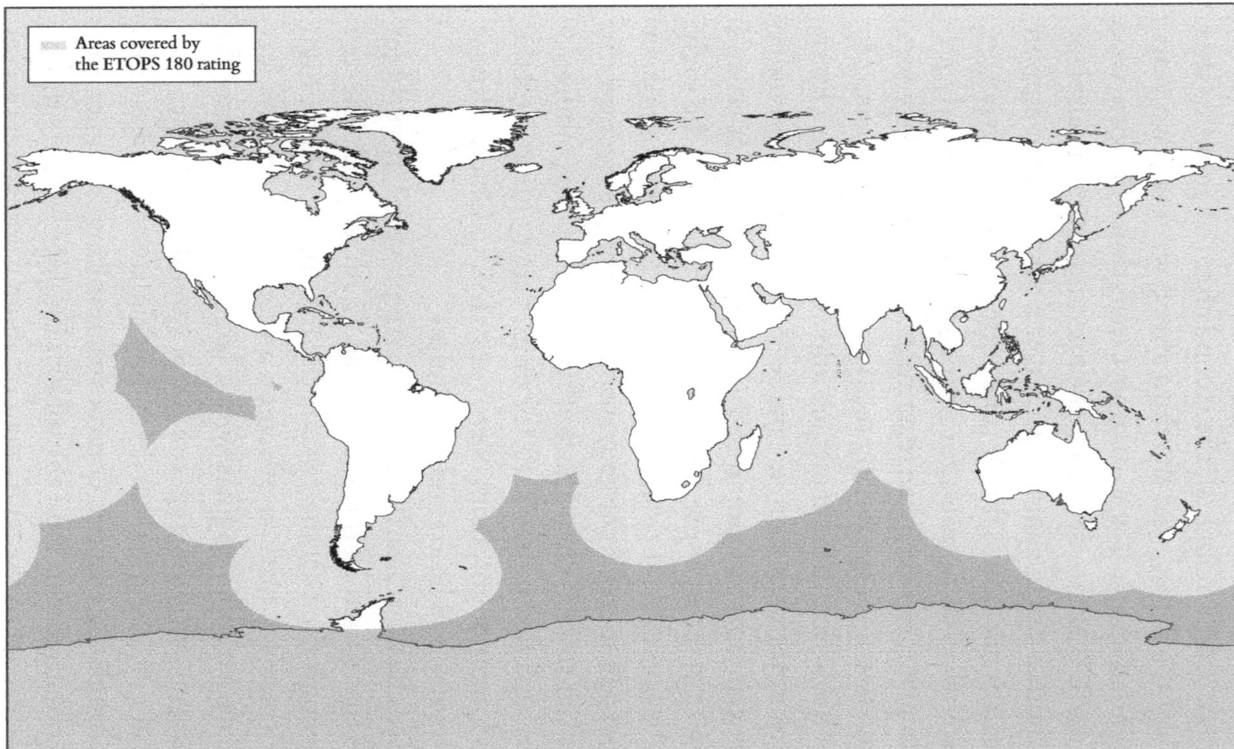

Areas that can be reached by adhering to 180-minute Twin-engine Operations Performance Standards

(to about 1,500–1,850 meters) and some business jets are pressurized to sea level even when cruising at 12 kilometers!

Modern long-range (twin-aisle) jetliners are wider and have larger windows than 50 years ago, and if airlines would not pack so many passengers in economy class (incredibly, Ryanair has even proposed strap-hanging, standing-only flights!) they could be roomier, with seat widths and leg room a bit closer to what is now the standard seating in business class. The most notable improvement in that class, now offered by many airlines, provides passengers on long-distance flights with fully flat beds as opposed to the only partially reclinable seats that were the old standard. And some first-class arrangements, now offered by a small number of carriers, even offer fully enclosed mini-cabins (obviously, for a maxi-price). As a result, in 2025 the difference between the cheapest economy and the most expensive first-class ticket on a one-way flight from Dubai to London could be as much as 20-fold!

First-class cabin on Emirates

The Long History and Uncertain Future of Electric Vehicles

In 1922 Henry Ford wrote in his autobiography how the Edison Company had objected to his experiments with internal combustion engines and how its bosses promised to hire him "only on the condition that I would give up my gas engine and devote myself to something really useful." During the 1890s, the early age of the car industry, theirs would not have been an irrational demand. In 1896, the first American track race (at Narragansett Park in Rhode Island) was decisively won by a Riker electric car, and in 1899 in France the bullet-shaped electric *La Jamais contente* (The Never Satisfied) was the first road vehicle to reach 100 km/h. Compared to the two alternatives—steam-powered and gasoline-powered cars—electrics were clean, quiet, safe, and easy to operate (no hot steam, no arm-wrenching cranking, no flammable liquids).

The Electric Carriage and Wagon Company began operating taxicabs in New York in 1897; in 1899, the US production of electric cars was, at 1,500 vehicles, 60 percent higher than that of gasoline-powered cars; and by 1901 Pope's Electric Vehicle Company was the largest maker as well as the largest owner and operator of motor vehicles in the country. Less than a decade later, it all unraveled. Pope's Electric went bankrupt in 1907, and Ford's Model T began its conquest of the country's car market in late 1908. For a century afterward,

Camille Jenatzy in *La Jamais contente* in 1899

electric vehicles were the perennial cars of the future, with some experimental designs appearing and rapidly disappearing without any impact.

Matters began to change only before the end of the 20th century, as concerns about automotive air pollution in California, soon joined by worries about global warming, led to increasing calls for zero-emission vehicles and for the decarbonization of the global energy supply—that is, for the eventual end to all fossil fuel combustion. In such cases as making steel or energizing jetliners or massive container ships, that was bound to be a lengthy and difficult process—both technically and financially—but there were no great technical obstacles to powering cars by electric motors, and the higher cost of such vehicles could be reduced by providing suitable government subsidies, helping a new industry to grow into a mature undertaking that could eventually decarbonize about 25 percent of the world's consumption of liquid fuels, which went to power passenger cars in the year 2000.

The shift began indirectly during the 1990s with the first hybrid electric vehicles (HEVs), which coupled an internal combustion engine with an electric motor to improve fuel economy and acceleration. This led to plug-in hybrid vehicles (PHEVs) with added lead-acid batteries, and then to GM's EV1 (1996–99) and Honda's EV Plus (1997–99), and to many designs of short-range (neighborhood) EVs. Finally, between 2008 and 2011 came the first highway-capable designs (Tesla Roadster, Mitsubishi i-MiEV, Nissan Leaf). A century after electric vehicles lost to internal combustion, the electric car industry was reborn, and this time it aimed to advance to its ultimate victory, doing away with all internal combustion engine vehicles (ICEVs).

As with most growth phenomena in their early stages, the numbers of EVs manufactured worldwide began to rise steeply. By 2012, the production of PHEVs was still well ahead of battery electric vehicles (BEVs; 61,000 vs. 59,000 units), but by 2015 the order was reversed (330,000 BEVs vs. 220,000 PHEVs), by 2020 the numbers were 2 million vs. 980,000, and in 2025 the gap grew further to 9.5 vs. 4.3 million—so nearly 14 million EVs. This growth was accompanied by many exaggerated forecasts of imminent EV dominance, including such patently ridiculous (and often-recited) claims made in 2017 (by a lecturer at Stanford University) that by 2025 all newly sold cars will be BEVs and that they will cost no more than ICEVs.

The reality has been quite different. About 68 million cars were sold worldwide in 2023, including 10 million EVs and 4.2 million PHEVs, which means that ICEVs were still almost 80 percent of the global automotive market. And while the bestselling US gasoline car (Toyota Camry) sold for $27,000–$35,000, the bestselling US EV (Tesla Model Y) sold for $47,000–$56,000 or for 60–75 percent more. Demands for an early end of ICEVs have hit especially hard in Germany, the country where cars were invented during the 1880s and whose automotive industry has been a keystone of the country's economy. In October 2024, Volkswagen announced the closure of at least three factories, and Oliver Zipse, BMW's CEO, warned that the German plan to ban all gasoline cars by 2035 was unrealistic, arguing for a more flexible approach to green Europe.

Meanwhile, in Norway a third of cars on the road were electric by 2025, and BEVs made up 95 percent of new car purchases. But that is an exceptional case (100 percent cheap electricity from hydro stations, and massive BEV subsidies made possible by the country's record earnings from oil and gas sales, perhaps the ultimate "green" hypocrisy). Looking past that unique exception, every realistic long-term forecast indicates that BEVs have a long way to go to displace all the world's ICEVs: in 2025 there were 1.4 billion gasoline- or diesel-fueled cars on the road, compared to the global fleet of just over 70 million EVs (just 5 percent of the total). The world has gone from zero (in 2000) to 5 percent of EVs in 25 years: what are the chances that it will go from 5 percent to 100 percent during the next 25 years?

That transition is not just about swapping the prime movers—fuel engines for electric motors—it will also require the massive mobilization of materials (a BEV needs three to four times as much copper as an ICEV), the construction of extensive charging infrastructure, securing new sources of reliable zero-carbon electricity supply, and developing new ways to recycle masses of spent batteries. The International

Energy Agency expects that in 2035 BEVs will make up half of all car sales worldwide, while a detailed modeling for the US shows that 65 percent of new passenger cars by 2050 will be electric vehicles. Clearly, an all-BEV world by 2050 is highly unlikely, and then there are the more than 100 million trucks and heavy offroad machinery (agricultural, mining, construction) that are much harder to electrify. Do not make a category mistake: swapping internal combustion engines for electric motors is not as easy as switching from landlines to mobile phones.

Why It's Not Just Foreign Wages that Make Faraway-Made Products Cheaper

You may live in San Francisco or in Sheffield, in Ontario or in Bavaria, but this you share: your underwear may be from Bangladesh, your socks from Pakistan, your shirt from China, your pants from Vietnam, your mobile phone from South Korea, you have just bought a mango from Brazil, oranges from South Africa, and blueberries from Peru, next week you will buy olive oil from Tunisia and a piece of IKEA furniture made in Indonesia. Citizens of affluent countries have been a full beneficiary of globalization, an old economic process that accelerated after 1990, reached its apogee during the second decade of the 21st century, and may now be weakening but will not disappear. Globalization is a multifactorial phenomenon *par excellence*, but as for getting item A to a destination B—and doing so on scales required by mass consumption, within a reasonable timeframe and at an affordable price—one variable obviously stands out: the cost of transport (COT).

Biologists use this term when referring to energy that organisms must spend per unit of their mass on various forms of locomotion per unit distance (in scientific units: joules/kilograms/meters). Swimming fish (buoyed by water) have the lowest cost of transport, flying is less expensive than running, and surface swimming (think of ducks or people) is the most energy-demanding mode of animal locomotion. In moving goods, the fundamental formula to measure COT is the same but the units are larger: given the typical distances covered, the most memorable way might be to compare energy costs in megajoules per ton of cargo per 1,000 kilometers (MJ/t/1,000 km). Comparative energy-cost calculations are easy to make once one knows the fuel consumption of trucks, trains, ships, and planes.

Many people are aware of some of the basic variables governing road transport by trucks. When fully loaded (about 25 tons) and traveling at typical highway speeds (80–100 km/h), America's dominant 18-wheeler trucks consume about 340 liters of diesel fuel per 1,000 kilometers. That (with an energy density of 36 MJ/L) translates to 12,240 MJ or about 500 MJ/ton. A diesel-powered train carrying 100 double-stacked containers (each one loaded with 14 tons of goods) needs about 0.6 liters for every ton-kilometer. As its total weight will be 2,200 tons (1,400 tons of cargo and 800 tons of locomotives and cars), it will consume 1,320 liters of diesel fuel or about 35 MJ/t moved over the distance of 1,000 kilometers.

Calculations for a fully loaded Boeing 747 cargo plane (maximum capacity of 113 tons and consuming 12 liters of kerosene per kilometer) result in an energy cost of about 3,700 MJ/t. In contrast, for a container ship carrying 10,000 standard steel boxes (each one containing 14 tons of goods) and consuming only about 0.015 kilograms of heavy fuel oil (40 MJ/kg) per ton-kilometer, the rate is just 0.6 MJ/t: a train needs 67 times more energy, a fully loaded truck more than 800 times as much, and a jumbo cargo plane more than 6,000 times as much. The comparison is overwhelming: the low energy requirement of containerized shipping (thanks to the world's largest and most efficient diesel engines) combined with the increasing size of vessels (the largest ones now carry more than

24,000 standard containers—see "An Underappreciated Technical Revolution," p. 116) has eliminated transportation cost as a variable of concern.

Here is what it means in financial costs. In early 2025 it cost $2,500 to ship a container from Shanghai to Hamburg. A standard 20-foot-long steel container can pack 70,000 fine shirts (and more than 80,000 T-shirts), and a good shirt will sell in Germany for at least $50. This means that in 2025, shipping a shirt from China to Europe cost less than a tenth of a percent (a mere 0.07 percent) of the shirt's price. That is an utterly insignificant charge, and it does not fundamentally change even after adding the cost of rail and/or truck transport from a port to the final retail destination. Much as modern electronics has eliminated distance in communication (you do not have to pay more for sending an email from Chicago to New Zealand than sending it next door to Milwaukee), the combination of modern ocean-going vessels, freight trains, and trucks has eliminated distance as a variable of concern in transporting goods: only if a retailer were selling a shirt for $2.50 would the cost of intercontinental shipping surpass 10 percent of its store price, still well within the range of easy price adjustments.

This reality also demonstrates why it will be very difficult to reverse the long-running trend of offshoring labor-intensive manufacturing and "reshore" those activities. When distance does not matter, an item for export can be made anywhere if that country has a sufficient labor force willing to work for low wages in tiring, repetitive jobs, and often under dismal conditions. Clearly that is not the reality in high-income countries with stagnant or declining populations and high rates of not only secondary but also postsecondary education, where manufacturing jobs are not even the third choice of most people entering the labor force.

Transferring a standard steel shipping container

And that is why the origins of manufactured items shift to places that still have higher fertilities and more people willing to endure the repetitive labor of manufacturing. The transfer of such labor from affluent countries began after 1950, with Japan emerging as the first distant major exporter of textiles, apparel, toys, and electronics; by 1970 the manufacturing was shifting into Taiwan and South Korea, after 1990 massively to China, and then from China to Vietnam, the Philippines, and Cambodia, as well as to Bangladesh, India, and Pakistan—and most recently, and despite its endless turmoil, to Africa (Ethiopia, Kenya, Lesotho). What comes next?

Wealth, Work, and the World

The First Age of Globalization

Just after he attended the Paris Peace Conference in 1919, John Maynard Keynes, at that time at Cambridge and later one of the 20th century's most influential economists, published *The Economic Consequences of the Peace* in which he recalled what had been possible in prewar London. To be sure, his was a privileged perspective (made clear by writing that "he could despatch his servant to the neighbouring office of a bank for such supply of the precious metals as might seem convenient") but his facts and conclusion captured the unprecedented realities of the era:

> What an extraordinary episode in the economic progress of man that age was which came to an end in August 1914! The greater part of the population, it is true, worked hard and lived at a low standard of comfort . . . but . . . for any man of capacity or character at all exceeding the average . . . life offered, at a low cost and with the least trouble, conveniences, comforts, and amenities beyond the compass of the richest and most powerful monarchs of other ages . . . The inhabitant of London could order by telephone, sipping his morning tea in bed, the various products of the whole earth . . . and reasonably expect their early delivery upon his doorstep . . . He could secure . . . cheap and comfortable means of transit to any country . . . But, most important of all, he regarded this . . . as normal, certain, and permanent, except in the direction of further improvement . . .

These realities were made possible by the intense and unprecedented period of economic globalization that took place between 1850 and 1914 and was based on the exploitation of new forms of energy, including mass-produced coal, crude oil (since the 1860s), and electricity (starting in 1882), which resulted in more than quintupling the global energy supply. Abundant energy and technical advances (led by the invention of internal combustion engines, electricity generation, electric lights and motors, and steel and aluminum production) increased the supply of new materials—above all of affordable steel (whose output rose more than 60-fold between 1850 and 1913) and Portland cement—and introduced new modes of manufacturing (driven by rapid industrialization) and new machines, whose adoption transformed transportation (advancing from steam-powered trains to the first electric locomotives, from slow paddlewheel steamers to large and fast ocean liners) and communication (progressing from telegraph to telephone and radio).

This led to unprecedented scales of mass production (culminating in the introduction of moving assembly lines in 1908), high levels of urbanization (globally from 10 percent to 20 percent), and intercontinental mobility. Some 60 million people emigrated worldwide between 1850 and 1913, with the largest streams including 30 million Europeans moving to the US and Canada, the British settling of Australia and New Zealand, flows of poor Italians from Calabria and Sicily to Argentina and Brazil, migrations from European Russia to Central Asia and the Far East, and the movement of Chinese and Indian laborers to the Americas and Africa. Quality of life improved thanks to rising incomes, better diets, and improved hygiene (bringing gains in longevity—US life expectancy for males rose from 40 to 58 years), and the expansion of foreign trade created the first era of economic globalization.

We have reasonably accurate aggregate economic data that make it possible to quantify the era's accomplishments. The first globalization began with falling trade prices in Europe in the 1840s: aggregate export openness (the ratio of total exports to GDP) in European trade had doubled between 1840 and 1870, and subsequent technical advances helped to expand worldwide trade during the next decades. Between 1850 and 1913, per capita GDP nearly doubled in the UK and Russia, in France it more than doubled, in the US and Germany it rose about 2.5 times, and in Japan it nearly tripled; meanwhile the world's per capita GDP more than doubled. The absolute value of foreign trade (exports and imports) rose about 2.5 times in Germany, doubled in the UK and Russia, and grew more than fivefold in the US.

And, most notably, between 1850 and 1913 the share of exports in total economic products of the then most industrialized countries increased by 50 percent, reaching more than 25 percent in the UK, Australia, and Canada and about 20 percent in Germany and Italy: remarkably, this made these economies almost as dependent on foreign trade as they were by the year 2000! The best global estimates indicate that in 1820 world trade contributed no more than 1 percent to the global gross economic product, that by 1850 the share had reached about 4 percent, and by 1913 had risen to 9 percent. That share may seem low in comparison with the peak level of post-1990 globalization (at nearly 25 percent before 2020) but the pre–First World War share (following the steep post-1914 decline and fluctuations of the following four decades) was matched only by the mid-1960s!

Prominent examples of pre-1914 trade on large intercontinental scales include shipments of energy and raw materials—coal (from the UK, US, and Germany), crude oil (from the US and Russia), lumber (from the US, Canada, Scandinavia, and Southeast Asia), rubber (from Brazil and Southeast Asia), cotton (from India, Egypt, the US), and Chilean nitrates (the first inorganic nitrogen fertilizer)—as well as staples (wheat from Russia, the US, Canada, Australia, and Argentina), chilled or frozen meat (from North and South America, Australia, and New Zealand), sugar (from Brazil and the Caribbean), and coffee (from Brazil, the Caribbean, East Africa), tea (from China and India), and spices (the trade dominated by Southeast Asia).

Other signs of globalization ranged from the extension of access to previously hard-to-reach destinations to the increased availability of affordable manufactured items that made much-appreciated gifts of toys or small luxuries. Access improved thanks to steamers deployed on rivers (be it the Congo, Nile, Niger, Amazon, or Yangzi) and to narrow-gauge railways climbing to high-lying settlements (be they in Asia or Latin America), and making the pilgrimage (Hajj) to Mecca became much easier as steamers departed from many Asian and African ports for Jeddah and a railway (completed in 1908) crossed the desert from Syria to Madinah. And among some 25,000 items listed during the 1890s in the Montgomery Ward and Sears, Roebuck & Company catalogs were many affordable German dolls ("bisque head, flowing hair, and long plaited chemise"), French perfumes, and creped Japanese paper napkins: more than a century before Internet shopping, even America's Great Plains farmers, far away from coastal cities, could partake in the first great wave of globalization.

Costliest Exports: From Diamonds to Glass Eels

The high intensity of trade has been one of the distinguishing features of modern economies. After the first wave of globalization was cut short by the First World War (see "The First Age of Globalization," p. 137), the second wave, which commenced shortly after the Second World War, now appears to have reached its peak during the second decade of the 21st century. Worldwide trade in resources, products, and services accounted for 31 percent of the global economic product in 2022 and for 29 percent in 2023, with the trade openness index (the sum of exports and imports of all goods and services, divided by gross domestic product and expressed as a percentage) being especially high in some European nations. In Germany, with its extensive high-value manufacturing, it is at about 100 percent, but similarly high levels now apply in some low-income countries that are exporting their valuable resources: perhaps most notably, cobalt and copper sales raise DR Congo's index to 95 percent.

When measured in terms of traded mass, global exports (as expected) are dominated by fuels and ores. About 3.4 billion tons of crude oil and the liquid fuels refined from it are now traded every year, worth about $2.5 trillion or nearly 11 percent of global trade. Global iron ore and coal exports amount to about 1.5 billion tons each. The second-most massive export category (with hundreds of millions of tons sold worldwide every year) includes cereals (mainly wheat, rice, and corn, adding up to about 500 million tons), steel (440–450 million tons in the early 2020s), bauxite (the world's largest source of aluminum, 150 million tons), and cement (110 million tons). These commodities dominate global exports by mass because of our indispensable needs for energy, food, and construction and industrial materials. At the opposite end of the spectrum are products that weigh very little but whose values per unit weight are many orders of magnitude higher than those of truly existential imports.

Due to their rarity, diamonds of exceptional quality—those that achieve the highest grades for the 4 Cs (exquisite cut, color and clarity, and exceptional weight measured in carats)—remain the most expensive exports per unit of weight. The priceless (as all diamond appraisals agree) Koh-i-Noor diamond aside (after frequent changes of ownership it ended up in the crown of Queen Elizabeth the Queen Mother and is now displayed in the Tower of London), the Cullinan, the world's second-most expensive diamond, is now valued at $400 million or about $640,000 per gram when using its original uncut weight of 3,106 carats (621.2 grams). Its value per unit mass would be nearly 1.3 billion times higher than that of crude oil (recently selling at about 50 cents per kilogram). Of course, gems in this category are exceptional and infrequent buys, not commodities for regular sale.

Looking at the broader gemstone market from the top, high-quality diamonds surpass $200,000 per gram; rubies, emeralds, and sapphires are generally an order of magnitude cheaper, some selling for more than $20,000 per gram. Common gemstones are much cheaper, with topaz and garnet going for up to $500 per gram. Compared to this, the sums spent on the rarest foodstuffs by foodies unconcerned about the cost are comparatively cheap. The best Madagascar vanilla beans are still less than a dollar per gram, the infamous Kopi Luwak coffee (with beans eaten, partially digested, and excreted by Asian civet cats before getting ground to make a cup) is no more than $1.20 per gram.

In Tokyo's top sushi restaurants, the fattiest cuts of bluefin tuna can cost $4 per gram, and elvers (tiny translucent baby "glass" eels shipped from Canada to Japan, where are they fed to maturity) sell for up to $5 per gram. White truffles from Alba (in Italian Piedmont) are between $5 and $10 per gram, a bit more than the saliva-joined swallows' nests gathered from the caves of Malaysia. True saffron (the dark red stigma and styles of crocus flowers) can be had for $5 and no more than $10 per gram (a gram is gathered from about 150 flowers). In January 2026 Kiyoshi Kimura, president of the Sushi-Zanmai restaurant chain, bought in an auction (as a promotion shtick) a 234-kilogram bluefin tuna for $3.2 million, and that works out to about $14 per gram for the whole fish and 5–6 times as much for its best edible parts (*otoro* and *chutoro*, the fattiest and medium-fatty parts of tuna belly). At that level, we are at, or above, the rate for beluga caviar (from albino sturgeons) that goes for up to $40 per gram.

These relative values are reflected in globally traded aggregates. The worldwide trade of worked (but not mounted) diamonds is now about $40 billion a year (equal to the recent annual GDPs of Nepal or Estonia), and they are the world's 38th-most traded commodity by value. In contrast, saffron, caviar, and truffles ranked recently between 3,600th and 4,000th, with recent global annual sales of $220 million to $240 million; and bluefin tuna sales are less than $150 million a year. Even such a comparatively rare fruit as persimmons (exported mostly by Spain and Azerbaijan) earns its producers $500

A 234-kilogram tuna sold for a record price in January 2026

million a year, and the global export of fresh apples (dominated by China, Italy, and the US) has been recently worth about $7 billion.

Obviously, there is a high profit in selling rarities and frivolities, be it gems or food, but to rank high in cumulative global trade while selling mostly water, nothing beats exporting water-rich fruits. These mostly small but enticingly packaged and beautifully colored natural containers are composed mostly (85–95 percent) of sweetish water with a bit of roughage and vitamin C. Worldwide strawberry (92 percent water) sales (led by Mexico, Spain, and the US) are now worth about $4 billion a year, while the sales of fresh or chilled tomatoes (with Mexico, the Netherlands, and Spain being their export leaders) have surpassed $10 billion. For those who wonder if I have not just made a mistake: no, botanically, the tomato, like apples, persimmons, or strawberries, is a fruit!

Why Do Some Countries Get More Investment than Others?

A company can invest in a foreign country and hope to make profit without getting directly involved. Or it can own (or co-own) a company in a foreign country, be responsible for its operation, and transfer its proven techniques, skills, and management procedures. That is the difference between portfolio investment and foreign direct investment (FDI). Obviously, FDI can make plenty of difference by creating new jobs and revenues (and new tax sources), and by introducing entirely new enterprises and products. FDI has been around for centuries, but it really took off only after 1960. Soon people could buy American fast foods made by American companies in Europe and Asia, or Japanese cars assembled by local workers employed by Japanese companies in North America and Europe.

For a long time, the US was by far the largest destination of FDI, and it remained the leader during the first two decades of the 21st century, when cumulative foreign investment in the US (led by Japanese, Canadian, and German companies) reached about $4.2 trillion. But half of the time since 2010, China, not the US, has been the world's largest FDI recipient. This is a new phenomenon. Foreign investment in China remained limited during the early 20th century, when Shanghai became a commercial capital of post-dynastic

China—and then came decades of Chinese civil war, war against Japan (1937–45), and the establishment of Communist Party rule (People's Republic of China) in 1949 that led to almost complete economic isolation from the West (and after the Khrushchev–Mao rift, even a break with the USSR). Mao died in 1976, and by late 1978 Deng Xiaoping began to enact (at first very slowly) his new open-door policy aimed at modernizing the country's backward economy.

In April 1980, the Beijing Air Catering Company (preparing flight meals for airlines) became the first Sino-foreign joint venture established according to the new opening policy. But the progress remained slow during the 1980s, and substantial investment inflows to China began only in the early 1990s and then they rose rapidly. Already by 1993 China had become the leader among the low-income countries in attracting FDI, and in 2002 it surpassed, for the first time, the US as the world's leading destination of global FDI. This had profound impacts on China's employment, industrial output, and exports. By 2010 just over 55 million workers (16 percent of the urban workforce) were employed by FDI enterprises, whose share of industrial output rose from just 2.3 percent in 1990 to almost 36 percent in 2003—and it was still above 25 percent a decade later. And in 2005, enterprises built with FDI accounted for over 58 percent of exports and nearly 60 percent of imports. By 2015 the country had approved more than 800,000 foreign investment projects and cumulative FDI had reached nearly $1.5 trillion.

As the domestic economy developed, shares of FDI in its overall activity inevitably declined, but even by 2020 foreign-invested enterprises still accounted for just over 40 percent of the country's exports and imports. In cumulative terms, between 1990 and 2020 China received $2.78 trillion of

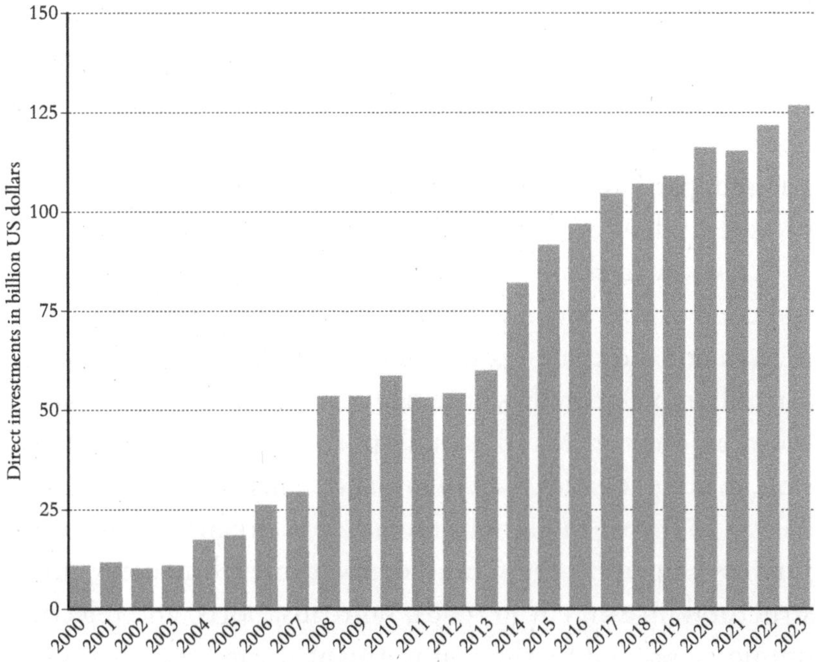

US foreign direct investment (FDI) in China, 2000–2023

FDI—almost exactly two-thirds of the monies invested during that period by foreigners in the US. That put the country far ahead of India (a total of $500 billion), as well of the two most populous countries in Latin America and Africa (Brazil, $640 billion; Nigeria, just $100 billion). What explains the discrepancy (which remains large even after adjusting for population size)?

Certainly not labor cost: scores of countries around the world offer much cheaper labor. There has been the obvious lure of skilled (literate and disciplined) labor available at mass scale and the promise of the world's largest consumer market—but India's population total has not been far behind and since 2023 it is the world's largest (and India is, finally, attracting much more FDI to make everything from apparel to mobile phones). Domestic economic policies have made an obvious difference. Once China changed its autarkic Maoist tune beholden to Stalinist economic precepts, its economy opened to the world much faster than anybody would have guessed, particularly in 1989 after the Party had violently suppressed student calls for greater openness.

Contrary to common expectations, the tightening of political control was accompanied by greater openness to foreign investment and trade—and this strategic decision led to a historically rapid growth whose rate matched and even surpassed the previous records of Japan's post-1950 economic expansion, making it easier than in just about any other low-income nation to invest in China's remarkable expansion. This was further helped by the Chinese government's massive infrastructural investment in roads, railroads, ports, and telecommunications, and by the commitment to an adequate energy supply, be it from domestic extraction and generation or from much increased oil and gas imports.

Neither India nor Nigeria has been able to match those policies. In comparison, by the year 2000 India still had an inferior infrastructure and less flexible bureaucracy. But, most fundamentally, it has been a matter of control—that is, of governance. In contrast to the inherent uncertainties of electoral democracies, foreign investors have welcomed the monopolistic rule by the Communist Party that offers reassuring predictability, stability, and continuity, unlike India's endless political contests and Nigeria's chronic lack of even temporarily effective governments. Not that China holds an exalted position: among 210 countries ranked by political stability (100 = stable) its percentile rank was just 25, but that is ahead of India's 21 and far ahead of Nigeria's less than 8.

Highly centralized governance also minimizes the risks of

surprising regional upheavals, recurrent internal tensions, or conflicts arising from regional, religious, and linguistic differences (these are always present in India and Nigeria), and the absence of labor unions makes for predictable labor force relations. And while China's one-party rule certainly does not preclude a great deal of corruption, India and especially Nigeria do far worse on this account: according to the 2023 corruption perception index, China ranks 76th out of 180 countries (worse than Jamaica or Saudi Arabia), but India is 93rd and Nigeria 145th. In any case, if you want to understand where national fortunes are headed you must take a close look at FDI and its trends.

The Most Important Variable for Understanding Stalling Growth

You cannot look deeper into the matter of economic growth without running into total factor productivity (TFP). Productivity is the simple part, being a ratio of outputs to inputs—telling us how efficiently the inputs are used. The term "total factor" comes from studies conducted by several US economists, most notably by Robert Solow at the Massachusetts Institute of Technology during the 1940s and 1950s. TFP is the share of economic growth that cannot be explained by accounting for labor and capital inputs, and that arises from advances and innovations in production techniques. In the long run, increasing TFP is the only way to sustain growth in per capita income (that is, improvements in the standard of living).

No economy can raise those standards simply by enlarging its labor force (that would, obviously, reduce overall per capita growth), and capital investments have diminishing returns: throwing more money at a problem has declining rewards, or it simply leads nowhere—to use limited resources requires their more efficient use, and such solutions require continuous innovation. Or as Solow put it in his 1987 Nobel Prize lecture, "the permanent rate of growth of output per unit of labor input . . . depends entirely on the rate of technological progress in the broadest sense." So how have we done in the past? Are we now living in an era of unprecedented technical advances that translate to soaring TFP? Will TFP growth remain high?

The answers are surprising. The historical growth of TFP is best followed by the growth of the US economy, and it shows notable fluctuations. TFP growth went from about 2 percent during the 1920s to 3 percent a decade later, remained high (2.5 percent) during the 1940s, and then slowed down to 2 percent between 1950 and 1973 (the year of OPEC's first round of crude oil price increases). Obvious factors behind this rise include the electrification of industries and services, the mass adoption of internal combustion engines, and the development of synthetic chemicals and new modes of telecommunication. The rise of widespread computing began during the 1970s (the first microprocessor and email in 1971, the first microcomputer in 1976), it accelerated during the 1980s (with the widespread adoption of PCs), and during the 1990s it was joined by the rapid diffusion of mobile telephones, and access to the World Wide Web, browsers, and Internet searches. And yet during the 1990s and the first decade of the 21st century, TFP grew by no more than 1–1.5 percent.

Similarly, the annual TFP growth of US manufacturing, the sector that most directly benefited from the combination of technical advances, was high (2–2.5 percent) between 1900 and 1929 thanks to rapidly progressing electrification, the introduction of assembly lines, and the commercialization of new processes. The rate was halved during the Great Depression in the 1930s, recovered a bit thanks to wartime production, and had its best period (2.5–3 percent) between 1946 and 1973 thanks to advances in automation and transportation. This era also ended with the OPEC oil price rise, and between 1974 and 1995 (despite microchips and the rise of computing) the annual TFP grew by merely 0.5–1 percent. TFP growth bounced back for less than a decade between 1996 and 2004 (2.5–3 percent), but during the next 15 years (before Covid's onset in 2020) it fell back to just 0.5–1 percent.

Robert Solow noted this paradox in 1987 when he wrote that "you can see the computer age everywhere but in the productivity statistics." What has caused these notable slowdowns in the growth of productivity? I find the most convincing explanations in the work of Robert Gordon (at Northwestern University in Chicago). Gordon shows that the progress of the supposedly "New Economy" that benefits from the advances in computing and telecommunication, enabled by rising microprocessor capabilities, has been inferior to the great inventions of the late 19th century. He concluded that the second industrial revolution of 1870–1900 that brought us electricity, internal combustion engines, gains in public hygiene, telecommunications, and new forms of entertainment, as well as new energy sources and materials, has mattered far more than either the first revolution (1750–1830, driven by steam engines) or the third one that started in 1960 and is based on advances in electronic computing and telecommunication.

This second revolution enabled eight decades of rapid productivity growth between 1890 and 1973, and once the commercialization of its technical advances was largely accomplished, productivity growth (except for a brief rebound between 1996 and 2004) slowed down. I found this argument so congenial because a decade before Gordon published his book in 2016, I came to the same conclusions in my two books (published in 2005 and 2006) about creating and transforming the 20th century. By tracing the paths of technical advances, I concluded that the decades between 1860 and 1913 amounted to a unique historical phenomenon as they brought an unprecedented, and perhaps unrepeatable, concatenation of technical progress—including electricity, internal combustion engines, automobiles, powered flight, new materials, and new chemical syntheses.

I also agree with Gordon's argument that the 21st century's advances, centered as they have been on personal entertainment and communication, do not fundamentally change labor productivity: in fact, such addictive distractions may lower it. But has not everything changed with the post-2022 eruption of superlatives regarding the coming, and supposedly totally unprecedented, "transformation of everything"—driven by accelerating advances in artificial intelligence? I urge an abundance of caution regarding such coming AI miracles. I am not denying that major productivity payoffs are possible (even very likely) thanks to the widespread deployment of AI solutions, but we will have to wait for at least another decade to see how

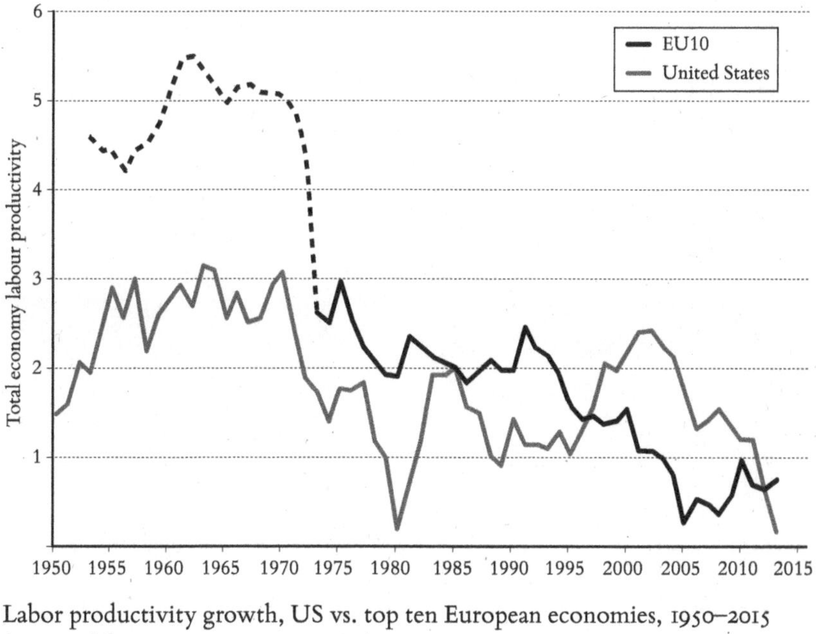

Labor productivity growth, US vs. top ten European economies, 1950–2015 (centered five-year moving averages)

stunning or disappointing that AI wave will have been. Long-term forecasts are always questionable, but the Congressional Budget Office has put future TFP growth at no more than 1.2 for the remainder of the 2020s and into the 2030s. If you want to understand where we are headed, watch for TFP.

Why the Black Economy Is Much Bigger than You Might Think

There is the economy traceable by national and international statistics that sum up raw materials extracted, food grown, products made, sold, and traded, services rendered, incomes reported, taxes collected, and pensions paid. Then there is the other economy that is not properly described by any of several commonly used adjectives: neither "shadow" or "underground" nor "black" or "illegal" economy will quite do because so many of these transactions are now done in the open, legally, and on scales that defy any oversight, control, regulation, or deterrence. At the apex are the leading banks, which have been caught setting up offshore accounts, hiding assets, and laundering money; at the bottom is the ageless barter trade in the poorest societies. And how many businesses simply underreport their income (or pay in cash), and how many street vendors, drivers, domestic workers, and casual laborers working openly in megacities or in crop fields around the world will never be registered—and their employers will pay no taxes and will evade all labor laws?

And these tax evasions are nothing compared to monies made through illegal activities including smuggling, human trafficking, prostitution, and drugs. All of these have always existed but have been perfected by using modern communication tools (a disposable mobile is a perfect enabler of fraud) and are now practiced with unprecedented intensity and with a truly global reach (Nigerian emailers were the great pioneers during the early years of the Internet, now think about semi-submersible boats transporting tons of cocaine from Colombia to Australia!), and they have been augmented by the worldwide trade in counterfeit goods (those $10 Versace shirts in an Asian market are obviously fake, but can you be sure that those expensive dried tomatoes in an Italian food store are organic?), cryptocurrency manipulations, telephone scams (you just won a free cruise, please, make a small handling deposit to this account: and too many people do!), and all kinds of cybercrime, ranging from hacking or denial of service for ransom (hospitals are a great target, as all patient information is now electronic) to selling stolen data.

Obviously, economists (and police forces) would like to know, at least approximately, the size of this other economy—and we have some ways to make some approximations. A direct approach (interviewing or auditing the perpetrators) might work as a sociological exercise done without divulging the sources and on a small scale (a perfect topic for a daring PhD thesis about organized crime in a small Calabrian town or in the production of Mexican avocados). Indirect ways—above all looking at national account discrepancies (gaps arising from comparing legal earnings with nationwide expenditures), analyzing trends in cash usage, and using electricity consumption as a reliable proxy for economic activity (does it rise much faster than official GDP?)—can be revealing. And, of course, economists love to build econometric models such as MIMIC (Multiple Indicators, Multiple Causes) that embrace many relevant variables but, as with all models, cannot do so without many questionable assumptions.

This means that we have plenty of figures of uncertain

value, and while some gross generalizations are undoubtedly correct (the relative size of the black economy tends to decline as societies get richer), closer looks may reveal sectoral or regional exceptions even in those cases. The most comprehensive global studies of the shadow economy have been done by Leandro Medina (economist at the International Monetary Fund in Washington) and Friedrich Schneider (economist at the Johannes Kepler University of Linz): they have published annual estimates of its size (as a percentage of official GDP) for more than 150 countries and, so far, extending from 1991 to 2017.

As expected, among the leaders are narco states (Bolivia at 55 percent), poor sub-Saharan countries (Nigeria 54 percent, Tanzania 47 percent), unraveled polities (Haiti 47 percent) and the former Soviet republics (Georgia 51 percent, Azerbaijan 49 percent). The most populous nations ranked from 11 percent for China and 19 percent for India to 34 percent for Brazil, while the richest economies scored between less than 10 percent (Netherlands, New Zealand, Japan, Canada) to 10–12 percent (Germany, France). The US (5.7 percent in 2017) and Switzerland (5.4 percent) had the lowest rates. Averages were about 14 percent for affluent economies, 28 percent for middle-income states, and 36 percent for low-income countries. How questionable these estimates are is best illustrated by a comparison of the US rate of 5.7 percent with the mean and state ranges calculated using the MIMIC model in a study spanning the period between 1999 and 2019 and published in 2023.

Its results show the informal economy averages 13 percent of the US GDP (more than twice the IMF's nationwide rate), with a minimum of 5 percent in Wyoming and a maximum of 21 percent in Georgia. In per capita terms the average (expressed in 2012 monies) was nearly $6,100—ranging from $2,924.49

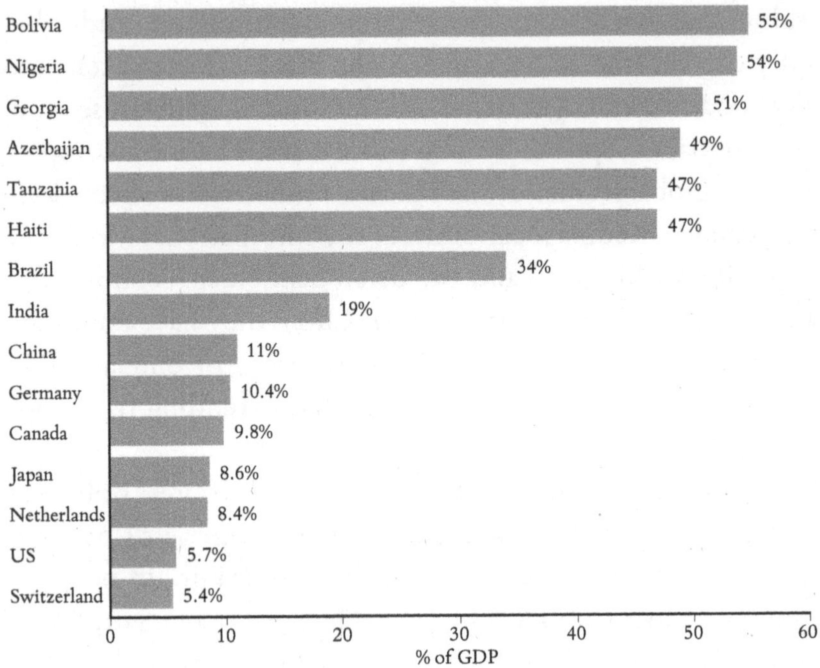

The black economy as a share of national GDPs

in Montana to $11,279 in Delaware. The American per capita black economy mean of some $8,500 (in 2025 monies) is hardly a trivial amount, as it is higher than the official GDP of more than 100 countries (Brazil is at about $9,000, South Africa at $7,000, and Indonesia at $5,000). On the other hand, extensive black economies mean that Bolivia, Peru, and Colombia are richer than the official averages indicate, and Spain's substantial black economy (20–25 percent of GDP) explains how the country coped in the past with its high unemployment rate that peaked at more than 25 percent in 2013 and is now below 15 percent: many people found their jobs in the shadows of society.

Undoubtedly, the black economy has its benefits, but will we ever account for its true costs that go beyond the reduced government revenues, distorted price competition, and widespread exploitation of labor? Will we ever put a price on activities that facilitate crime, provide cover for physical and sexual abuse, and imperil and destroy lives, be they those of illegal migrants (paying dearly for rickety boats or for trans-border guides) or teenagers caught in the drug and prostitution trade?

What Use Is Complexity?

In 2009, Cesar A. Hidalgo (at that time at the Massachusetts Institute of Technology) and Ricardo Hausmann (at Harvard University) published a paper entitled "The Building Blocks of Economic Complexity." That was followed by the publication of *The Atlas of Economic Complexity* (in 2012) and by the establishment of the Observatory of Economic Complexity (OEC) at Harvard, which now provides detailed annual global trade data covering 5,000 products and thousands of companies. Readers interested in the math underlying the original complexity model can download it from the website maintained by the OEC.

For this essay no simplified explanation could surpass their original Lego analogy:

> We can create indirect measures of the capabilities available in a country by thinking of each capability as a building block or Lego piece. In this analogy, a product is equivalent to a Lego model, and a country is equivalent to a bucket of Legos. Countries will be able to make products for which they have all of the necessary capabilities, just like a child is able to produce a Lego model if the child's bucket contains all of the necessary Lego pieces.

To carry on with the analogy, could we then infer the diversity and exclusivity of the Lego pieces inside a child's bucket by looking only at the models made by children using different buckets of Legos?

The authors show that this is possible and that connections between countries and their exported products inform us of a specific country's capabilities—much as a particular Lego model made by a child informs us of the specific set of Lego pieces they had to choose from. The diversity of exports that a country can produce, their abundance, and the number of other countries that can do the same give us a comprehensive and complex insight into the country's capabilities and income, as well as its prospects. A country may be a major exporter, but what it sells might be sold by many other nations: nearly 40 countries export crude oil; nearly 30 countries are major apparel exporters, while more than 50 nations have smaller but significant clothes export industries. A country may have a large trade surplus and high quality of life, but its prosperity might rest on just a few commodities and hence its economic complexity is low.

That is why affluent Norway (nearly 80 percent of its export value comes from crude oil and natural gas) ranks rather low—43rd among 133 studied countries, below such poorer but more export-diversified economies as Bulgaria, Bosnia, and India. Similarly, Canada, a major exporter of all forms of energy (oil, gas, coal, uranium, hydroelectricity) is 31st, less economically diverse than Thailand or Mexico. Who are the leaders? Worldwide rankings of economic complexity have been calculated going back to 1998, and Japan has come out on top every year since the beginning of the 21st century. Other top entrants shift, with Switzerland, South Korea, Germany, and the Czech Republic remaining in the top ten.

All of them owe their position to highly diversified industrial production ("manufacturing" is a misleading term in the age of microchips and robotics). In Japan, exports amounting to more than 2 percent of annual value include cars, machinery, integrated circuits, and motor vehicle parts, and their aggregate share is only about 27 percent as there are scores of other exported categories. In Germany the top ranking belongs to cars, medicaments, car parts, and vaccines, whose aggregate value is just 20 percent of the total—leaving room for a large variety of products ranging from medical instruments to plastics and from optical fibers to hops for brewing beer. In contrast, just three commodities (crude oil, liquefied natural gas, and refined oil products) make up 88 percent of Nigerian exports, leaving the country with an economic complexity ranking of 127 among 133 nations. And who is at the bottom? Mali, a West African landlocked desert country whose largest export, gold (sold mostly to the United Arab Emirates, and also to Switzerland and Australia), accounts for 96 percent of its foreign earnings, followed by raw cotton (less than 2 percent).

But simple economies can develop, and some countries have shown steep trajectories of rising complexity. Between 2000 and 2022, the Philippines rose from 65th to 35th place worldwide, China from 41st to 18th, and, most notably, South Korea from 20th to 2nd! At the same time, the US has retreated from 6th to 14th place (a consequence of deindustrialization and manufacturing losses, mainly to China), while many countries showed no or merely marginal up or down changes in their ranking. And as with any comparative ranking, complexity based on trading patterns is only part of a much more complex story, and Canada is an excellent illustration of this reality.

Given its high dependence on exports of natural resources (oil, gas, coal), Canada was never as complex an economy as the US or Germany, but recent deindustrialization lowered the country's economic complexity ranking from 24th place in the year 2000 to 41st in 2022, one of the most pronounced retreats in the global record. But at the same time Canada's economic complexity index based on number of patents put

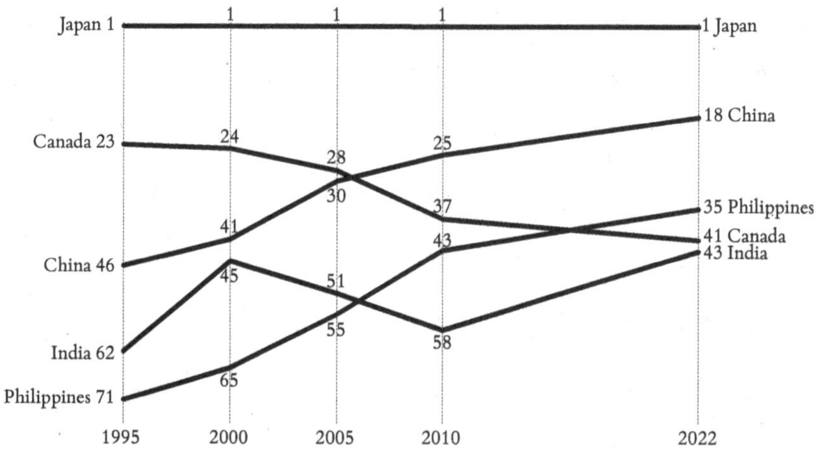

Changes in economic complexity ranking

the country at 11th place globally, and its economic complexity index based on scientific publications was the world's third highest behind the US and the UK. And how does economic complexity correlate with the Human Development Index (HDI), calculated as a geometric mean of life expectancy, years of schooling, and GDP per capita?

The highest HDI now belongs to Norway, followed by Switzerland and Australia—but, as already noted, Norway's economic complexity ranks only 43rd and Australia's is 78th, behind Guatemala whose HDI rank is 127! These are old comparative lessons: no composite index can sufficiently capture the complex realities of national economic and social development; there are always some notable outliers and remarkable paradoxes; and reducing everything to single numbers (even more so to indexes created from raw data) does not provide the best perspective or the wisest guidance. At best, it is an interesting league table.

Lessons from the Past, Warnings for the Future

The Rise and Fall of Portugal

History has no shortage of small countries, or small populations, that played, temporarily, outsized continental or even global roles. The Mongolian outburst from the region south of Lake Baikal all the way to central Europe (1206–1279), and the subsequent direct and indirect Mongol rule over vast portions of Eurasia, is perhaps the most obvious example: in 1227, when Genghis Khan died, the total population of Mongols (including the broader confederation of tribes under his rule) was only 1–2 million people, with mounted warriors amounting to no more than a tenth of that total. Europe offers several short-lived examples of such disproportionate reach, including the Vikings of the 9th to 11th centuries (whose influence reached from Norway to Sicily, and from the British Isles to the Black Sea) and early 17th-century Lithuania, when the Polish-Lithuanian Commonwealth ruled the lands extending from today's western Poland to eastern Ukraine and Russia's Smolensk region, and from today's Latvia to the borders of Romania.

Undeniably, Portugal is a bright star among this peculiar constellation.

Norse voyages to Iceland and Greenland aside, early 15th-century Europe's Atlantic horizons were as constrained as they were at the end of the Roman Empire, with Cape Bojador (latitude 26°N) being the southernmost place reached along the western coast of Africa. Portuguese sailors had discovered

Madeira by 1418, the Azores by 1427, and during the reign of Prince Henry the Navigator (he died in 1460), they used their caravels and carracks to explore the uncharted ocean surrounding Africa. In 1444 they reached the mouth of the Senegal River (about 16°N); in 1482 they crossed the equator and left rock inscriptions at the mouth of the Congo (5°S).

In 1488 Bartolomeu Dias rounded the southernmost shores of Africa and his ships entered the Indian Ocean, and on May 20, 1498, Vasco da Gama's ships landed in India after crossing the Indian Ocean from Africa to Asia. As immortalized by Luís Vaz de Camões in his epic poem *Os Lusíadas* (1572):

Já no largo Oceano navegavam,
As inquietas ondas apartando;
Os ventos brandamente respiravam,
Das naus as velas côncavas inchando.

In Richard Burton's classic 1880 translation: "They walked the water's vasty breadth of blue, parting the restless billows on their way; fair favouring breezes breathed soft and true, the bellying canvas bulging in their play."

This was an epoch-making change: one of the foundations of early modern history. The direct route from Europe to India—bypassing the land journey that had been used since antiquity to import Asian luxuries (spices, silk)—was the beginning of commercial globalization based on long-distance maritime trade, a reality that now defines the modern world. And the direct route to India was not the end of Portuguese sailings: by 1511 they reached the Straits of Malacca, by 1519 they began trading with the Spice Islands (Banda and the Moluccas) in the Indonesian archipelago, and by 1543 they reached Japan, providing the first commercial (and soon also cultural and religious) links between the easternmost Asian nation and Europe—which were maintained until the final expulsion of all Portuguese in 1639.

In the western hemisphere, Brazil was made officially a colony in 1549 but its development led to Portugal's most unenviable record: that of being the greatest slaving nation in history in absolute terms, and even more prominently in relative (per capita) terms. Between 1501 and 1866 Portuguese ships transported 5.85 million slaves, or some 47 percent of the nearly 12.5 million people abducted from Africa, while British ships carried 3.3 million, French 1.1 million, Spanish about 1.1 million, Dutch half a million, and US vessels about 300,000.

But neither the head start in maritime exploration nor the possession of far-flung colonies, massed slave labor, and the associated trade helped Portugal to retain its fleeting time of glory. Between 1580 and 1640 the country was ruled from Spain, and much like Spain it did not make any stunning advances during the next 200 years of its existence. We do not have any long-term reconstructions of Portugal's economic growth, but it is almost certain that its per capita GDP could not have done better than did the Spanish one that made no progress between 1400 and 1700! For comparison, during that period the Dutch per capita GDP doubled and the English one grew by nearly 60 percent.

The discovery of Brazilian gold (1690s) did not result in any major or lasting economic gain, and it was far outweighed by the loss of Brazil (in 1822) and the near absence of Portugal's industrialization during the 19th century. The first republic (1910) followed the constitutional monarchy, and the country's lagging performance was not helped by nearly four decades (1932–1968) of Salazar's dictatorship. The restoration of democracy (in 1974) and the granting of independence to Portugal's African colonies led to prolonged civil wars in Angola and Mozambique, resulting in a flood of returnees. Gradual improvements in quality of life accelerated only after the country joined the EU in 1986, but even after decades in the Union the country's key indicators remain below the group's mean: GDP per capita ranks 18th out of 27 member countries, and in terms of the Human Development Index (see "Can We Measure Happiness?" p. 27), Portugal is 21st.

Portugal's trajectory—rising rapidly to global prominence and quickly retreating to prolonged stagnation and economic marginality—is hardly unique. There is the glorious past captured by Camões—"You Portuguese are adding to the world, / Opening the gates to that vast ocean / Which you navigate with such courage . . ."—and no glorious aftermath. But the nation has endured: thanks to its location (as

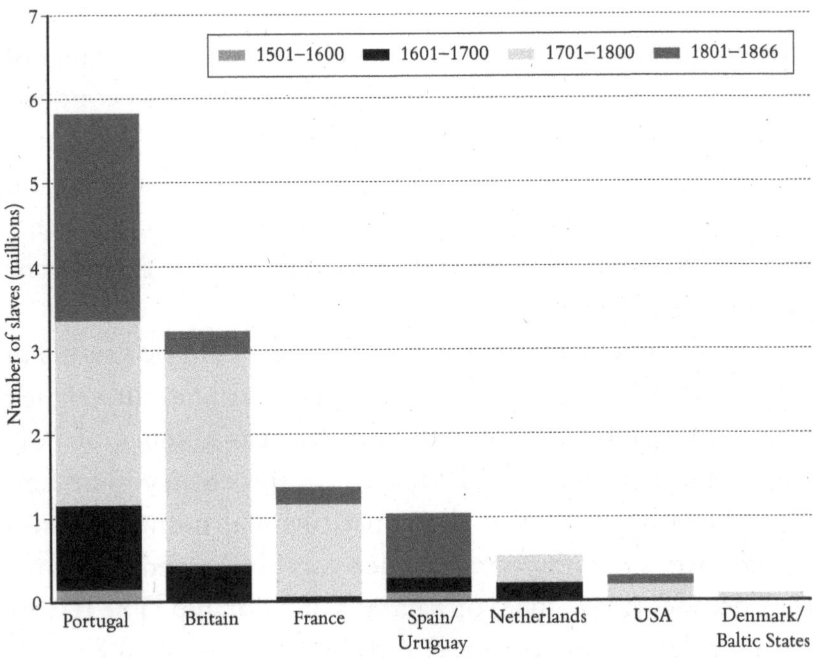

Estimated totals of slaves transported by the largest slaving nations, 1501–1866

the westernmost country of the Eurasian continent) and to the relative weakness of its only land neighbor, Portugal was able to preserve its sovereignty and identity—the quest that many formerly more powerful countries found historically much harder, or impossible, to do.

The History of Cranes (Machines, Not Birds)

Early in the second century CE, a well-off Roman couple engaged in the construction business were buried in a large tomb along the Via Labicana, southeast of Rome. The Tomb of the Haterii (discovered in 1840) contained busts of the deceased and, appropriately, a relief depicting the construction of a temple-like funerary monument that contains the oldest well-preserved illustration of a treadwheel-powered wooden crane, with five men walking inside the wheel and two perched on the crane's top. By that time, cranes—machines that can lift a burden, move it sideways, and lower it, and hence indispensable for lifting and shifting any heavy loads that must be incorporated in structures or transported by vessels—had been in use for more than half a millennium: the Greeks employed them when building marble temples 25 centuries ago.

The two fundamental differences between ancient cranes and modern versions involve the materials that make up those cranes and the energies that power them. The cranes of early Greek antiquity were just wooden jibs with a simple pulley whose mechanical advantage came from winding the ropes by winches. By the third century BCE, compound pulleys had come into use and the Roman *trispastos* (three-pulley cranes) provided a nearly threefold mechanical advantage: nearly, because there were losses to friction. Those losses became

The Tomb of the Haterii crane

prohibitive with the addition of more than the five ropes of the *pentaspastos* cranes.

The most powerful cranes in the Roman and later the medieval periods were powered by men treading inside wheels. In his 1563 chaotic painting of the Tower of Babel, Pieter Brueghel the Elder shows such a wooden crane lifting a slab of stone. Other cranes were operated by animals turning windlasses by walking in tight circles, and their lifting capacities were generally between 1 and 5 tons.

Major advances in lifting power came only during the 19th century. During the 1850s William Fairbairn, a Manchester engineer, built a large tower crane powered by four men turning winches. Its performance was further expanded by fixed and movable pulleys, which gave it more than a

600-fold mechanical advantage, enabling it to lift weights of up to 60 metric tons and move them over a circle with a 32-meter diameter. And by the early 1860s, William Armstrong's company was producing annually more than 100 hydraulic cranes (transmitting forces through liquid under pressure) for English docks. The steam-powered cranes of the latter half of the 19th century were able to lift more than 100 metric tons in steel mills; some of these cranes hung above the factory floors and moved on roof-mounted rails. During the 1890s, steam engines were supplanted by electric motors.

The next fundamental advance came after the Second World War, with Hans Liebherr's invention of a tower crane that could swing its loads horizontally and could be quickly assembled at a construction site. Its tower top is the fulcrum of a lever whose lifting (jib) arm is balanced by a counterweight, and the crane's capacity is enhanced by pulleys. Remarkably, when Liebherr presented his first mobile tower crane at the Frankfurt autumn trade fair in 1949, the response was disappointing: "Actually, when the trade fair was over, I could have brought my production in crane construction to an end." But that soon changed, as tower cranes greatly contributed to the reconstruction of bombed-out German cities and then diffused rapidly to construction and other projects.

By 2007 Liebherr, now a German Swiss multinational company, had become the world's largest producer of cranes, employing more than 50,000 people. Liebherr now produces many designs of large tower cranes in three major categories: fast-erecting cranes, top-slewing cranes (rotating around their vertical axis), and mobile cranes. The maximum lifting capacities of Liebherr's stationary tower cranes range between 6 and 40 metric tons, with tip capacities between 1.6 and 6.9 tons and a radius of up to 82 meters. As for the lifting capacity, the record is held by the K10000, made by the Danish firm Krøll Cranes. It can lift as much as 120 metric tons at a maximum radius of 100 meters. Liebherr also makes many mobile models, and its largest—a truly gargantuan mobile crane—sits on an 18-wheeler truck. It has a 100-meter-long telescopic boom, a maximum hoist height of 188 meters, and it can support 1,200 tons! Construction ring cranes do not have a single narrow central spindle as they move on a ring-shaped track, with the largest model, *Big Carl* made by the Belgian firm Sarens, having a 275-meter reach and a maximum lift of 5,000 tons.

Cranes are also mainstays of global commerce: without gantry quay cranes capable of hoisting from 40 to 80 metric tons, it would take weeks to unload the more than 20,000 standard-size steel containers that sit in one of today's

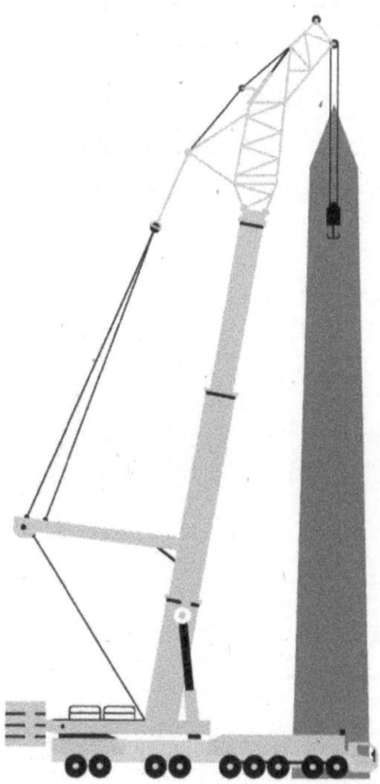

Liebherr's largest mobile crane compared with the Washington Monument

enormous container ships (see: "An Underappreciated Technical Revolution," p. 116). But put cranes together in a tightly coordinated operation involving straddle carriers and trucks, and the job now takes fewer than 72 hours. And in June 2017, 20,568 containers were unloaded from the *Madrid Maersk* in Antwerp in a record time of 59 hours. Without these giant cranes, every piece of clothing, every pair of shoes, every TV, every mobile phone imported from Asia to North America or Europe would take longer to arrive and cost more to buy.

But the world's most powerful crane was not built for construction on land or for loading vessels but for lifting heavy modules during the assembly of semisubmersible platforms used for offshore oil and gas drilling and for floating production storage and offloading projects: the Taisun crane works in Yantai Raffles Shipyard in China's Shandong province and can hoist 20,000 tons—or a burden about half again as heavy as the Brooklyn Bridge! The quotidian functioning of the modern world depends on many machines and processes, but every large-scale effort to build or move something substantial starts with cranes.

The Practical Limits of Design

In a world accustomed to the constant growth of aggregates (be it rising GDPs, higher exports, or expanding sales of electric cars) and performances (best exemplified by the soaring capabilities of electronic computing, data processing and storage, and mobile devices) it is salutary to be reminded of the fact that many designs that have made our civilization so prosperous reached their functional limits generations ago. The comparison of a single number—the modulus of elasticity (which measures the stiffness of elastic materials as they are stretched or compressed)—makes it clear why we do not have large-span wooden bridges (or large wooden jetliners). For construction steel, the modulus is about 15 times higher than for wood; for the carbon-fiber-reinforced polymer now used to make jetliners it is more than 40 times higher—next time you fly on a Boeing 787 get a window seat from which you can see a wing, and watch how it flexes during flight. Another measure that offers a fascinating historical perspective on design limits is the length-to-beam ratio (LBR) of ships.

This ratio is simply the quotient of a ship's length and breadth, both measured at the waterline; you can think of it as the expression of a vessel's sleekness. A high LBR (long slender hull) favors speed but restricts maneuverability as well as cargo hold and cabin design. These considerations, together with the properties of shipbuilders' materials, have always limited the LBR ratio of large vessels to single digits.

Here are the numbers in their historical sequence. The simplest floating object that can carry a load has been used since antiquity in locales ranging from Wales to Tibet: all you need is a rough wickerwork over which you stretch thick animal skins to get a man-size circular or slightly oval coracle that is good to use on rivers or lakes. Such a craft has an LBR close to 1, so it is no vessel fit for crossing an ocean, but in 1974 an adventurer did paddle one across the English Channel.

Building with wood allows such sleek designs as the canoes carved from single massive cedar trees by the Haida, Tlingit, and Nuu-chah-nulth of the Northwest Pacific coast: they had an LBR as high as 11–13 (about 20 meters long, 1.5–1.8 meters wide) but obviously they could not cross oceans or carry substantial cargo. The LBR of ancient and medieval commercial wooden sailing ships built from wood planks increased slowly, and can be reconstructed accurately from many extant wrecks. Roman vessels transporting wheat from Egypt to Italy had an LBR of about 3; ratios of 3.4–4.5 were typical for Viking ships, but their lower freeboard—the distance between the waterline and the main deck of a ship—and much smaller carrying capacity made them even less comfortable on long-distance voyages. The *Santa María*, a small carrack captained by Christopher Columbus in 1492, had an LBR of 3.45 (19 meters long, 5.5 meters wide). With high prows and poops, some small late-medieval carracks had a nearly semicircular profile. Caravels, used on the European voyages of discovery during the following two centuries, had similar dimensions, but multi-decked galleons were sleeker: the *Golden Hind*, which Francis Drake captained to circumnavigate the globe between 1577 and 1580, had an LBR of 5.1.

Little changed over the following 250 years. Packet sailing ships, the mainstays of European emigration to the United

States before the Civil War, had an LBR of less than 4. In 1851, Donald McKay, Boston's famous designer of sleek and fast clipper ships, crowned his career by launching the *Flying Cloud*, whose LBR of 5.4 (69 meters long, 12.7 meters wide) reached the practical limit of nonreinforced wood; going well beyond that ratio, loaded hulls in heavy seas would strain, bend, twist, and eventually break. But by McKay's time, wooden hulls were on the way out.

In 1845, SS *Great Britain* (designed by Isambard Kingdom Brunel, at that time the country's most famous engineer) was the first iron vessel to cross the Atlantic. Its LBR was 6.4 (98 meters long, 15.4 meters wide). Inexpensive steel (thanks to Bessemer process converters) induced Lloyd's of London to accept its use as an insurable material in 1877, and in 1881 the Cunard Line's SS *Servia*, the first large transatlantic steel-hulled liner, had an unprecedented LBR of 9.9 (157 meters long, 15.9 meters wide). And that was the limit—as the dimensions of future steel liners clustered close below that ratio: 9.6 (269.1 meters long, 28.1 meters wide) for RMS *Titanic* (launched in 1912); 9.3 for SS *United States* (launched in 1951); and 8.9 for SS *France* (launched in 1960, two years after the Boeing 707 began the rapid elimination of transatlantic passenger ships).

Huge container ships, today's most important commercial vessels (see "An Underappreciated Technical Revolution," p. 116) have relatively low LBRs to accommodate packed rows of standard steel container units. *MSC Irina* (launched in October 2022)—the world's largest, with a capacity of 24,300 container units—is 399.9 meters long and 61.3 meters wide; hence its LBR is only 6.5, the same as that of *MSC Gülsün*, the previous record holder. Royal Caribbean's *Icon of the Seas* (launched in July 2024)—the world's largest cruise ship, built in Finland and able to board up to 7,600 passengers and a crew

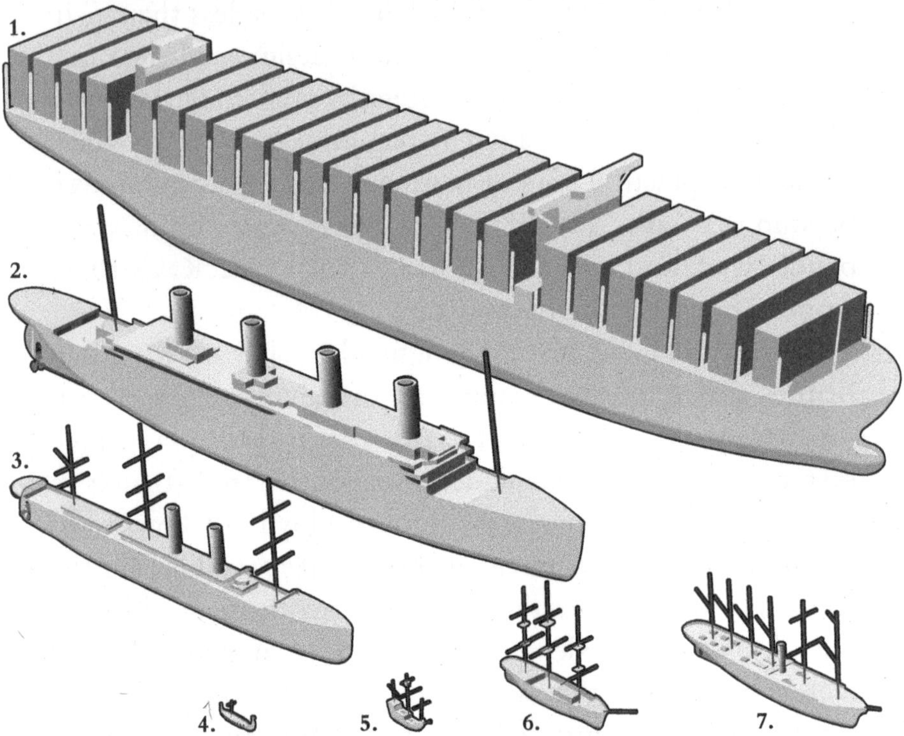

1. *MSC Gülsün*
LBR: 6.5

2. RMS *Titanic*
LBR: 9.6

3. SS *Servia*
LBR: 9.9

4. Viking longship
LBR: 4.4

5. *Santa María*
LBR: 3.45

6. *Flying Cloud*
LBR: 5.4

7. SS *Great Britain*
LBR: 6.4

Length-to-beam ratios

of 2,350—is less than 10 percent shorter (364.75 meters), but its narrower beam (48.47 meters) gives it an LBR of 7.5. And to complete the comparisons, USS *Gerald R. Ford*, the largest US aircraft carrier (launched in 2017), has an LBR of 8.2 (length 337 meters, beam at waterline 41 meters).

There are much sleeker vessels around, but they are designed for racing, not to carry massive loads of goods or passengers. Each demi-hull of a catamaran has an LBR of about

10–12, and in trimarans, whose center hull has no inherent stability (that feature is supplied by the outriggers), the LBR can exceed 17. Think about these limits when you read about AI boosting everything to unheard-of levels: it might do that to the gullibility of small-screen readers but not, among many other things, to LBR.

Icon of the Seas

Extreme Designs

There is a fundamental difference between what can be designed and built and what makes sense. History provides lessons in the shape of record-setting behemoths that have never since been equaled, and one of the most remarkable extremes is now more than 4,500 years old. The Egyptian pyramids started small, but in just a few generations there was Khufu's enormous pyramid (completed in the early 25th century BCE)—which nobody has ever tried to surpass. Of course, even higher stone heaps could be assembled by modern construction methods, but we could not replicate the original building process because we still do not know the exact construction techniques used to erect the great pyramid.

Oared vessels offer another example of an unequalled extreme design going back to antiquity. Shipbuilders in ancient Greece kept on expanding the size of their oared vessels until Ptolemy IV Philopator (221–204 BCE) ordered the construction of a *tessarakonteres*, a ship with 40 banks of oars, 4,000 oarsmen, and 400 sailors, able to carry nearly 3,000 warriors. But as Plutarch wrote nearly three centuries later, "this ship was merely for show; and . . . she was moved only with difficulty and danger." The vessel, too heavy and too ponderous, was a design failure that ended as a costly stationary display.

The modern era has no shortage of such obvious design and construction overshoots. The boom in oil consumption following the Second World War led to ever-larger oil tankers, with maximum sizes growing 10-fold between the early 1950s and early 1970s, and between 1976 and 1979 France built (in the Saint-Nazaire shipyards) four *Batillus*-class supertankers, each rated at 555,000 deadweight tons (dwt) and capable of carrying more than half a million tons of crude oil. Their domination was short-lived: three of these ships were scrapped after less than a decade in service; the fourth one was decommissioned in 2003. At 657,019 dwt, *Seawise Giant*, launched in 1979 in Japan, was the largest self-propelled vessel ever built. In 1988 the tanker was damaged by an airstrike during the Iran–Iraq War, repaired and renamed (*Jahre Viking*), converted to floating oil storage in 2004, and scrapped in 2010. Proposals for a million-dwt tanker published during the 1970s were never realized. Technically, building such a ship is possible, but it would not fit through either the Suez or Panama canal, its draft would limit its operation to just a few ports, and it would be prohibitively expensive to insure.

Flying offers two notable examples of unrepeatable excess. One of the most infamous overshoots of engineering design was the H-4 Hercules airplane designed during the Second World War by the Hughes Aircraft Company as a transatlantic strategic airlift flying boat to supply US armies in Europe. Due to wartime restrictions on aluminum use, it was built entirely of birch wood, and the only prototype, with a wingspan of nearly 100 meters and propelled by eight reciprocating engines, was completed only after the war ended. It became airborne only once, for less than a minute, on November 2, 1947, with Howard Hughes himself at the controls. Nearly eight decades later the plane (known as the *Spruce Goose*) remains in good shape, and it is displayed to the public at the Evergreen Aviation & Space Museum in McMinnville, Oregon, a reminder of excess in aeronautic design.

The other notable excess in aviation design came six decades later: the intention to build Airbus A380 was revealed in 1994, the first prototype flight took place in 2005, and the plane entered commercial service in October 2007. The economy-class-only configuration of the Airbus A380 airliner was certified to carry up to 853 passengers, and a typical two-class configuration was for 525 passengers—but the plane has not been a commercial success. The plane's production ended in 2021, just 11 years after it entered service and after the delivery of just 251 planes, a very truncated lifespan and a disappointing production run. Compare it with the plane that has held the capacity record since its launch in 1969, Boeing's 747: its cumulative deliveries reached 1,574 planes, with the last four orders (for cargo versions) made by Atlas Air in 2021, 52 years after the plane's first flight. Clearly, the design's longevity and commercial success prove that the famous jumbo jet was the right-sized long-lasting record-breaker.

Another right-sized American giant design has been Ford's heavy (up to 2.2 tons) and powerful (up to 537 kilowatts) F-150 pickup truck, now (since 2021) in its 14th generation. The first model F-150 went on sale in the US in 1975 and it became the country's bestselling pickup just two years later. Most remarkably, the F-150 has been America's bestselling road vehicle since 1981, now for more than four decades and counting. In contrast, the Hummer, a civilian version of a military assault vehicle (Humvee) weighing nearly 3.5 tons, had a limited commercial career (1992–2010), but in 2021 it was resurrected in an even heavier electric version weighing 4.5 tons.

But these lessons of excess carry little weight with designers and promoters pursuing record sizes. As the construction of the world's first kilometer-tall building (Jeddah Tower) resumed (after a five-year pause) in 2023, structural engineers are now thinking about buildings taller than a mile (1,609 meters). Cruise ship designers who have already packed up to 7,600 passengers and people are dreaming about much larger (and permanently) floating cities. There are also engineers who think that we will soon have wind turbines whose blades will be more than 200 meters in diameter and that they will fold, like palm fronds, to prevent damage in hurricanes.

Depending on where you stand, you might see all of this either as an admirable inventive quest for scaling new heights in design (quintessential examples of human striving) or as an irrational and wasteful overreach (recurrent demonstrations of human hubris).

Airbus 380

Why Gradual Improvements Trump Eureka Moments

The popular view of technical progress sees innovation primarily as the achievements of singular heroic inventors: they come up with brilliant new ideas and change the world. James Watt patents a steam engine and the world gets the industrial revolution. Thomas Edison comes up with a durable light bulb and banishes the darkness. Carl Benz builds a small gasoline-fueled internal combustion engine and a century later the world has a billion cars . . . etc., etc. This is incorrect, naive, and misleading. None of those men came up with the principal idea: they made better, practical versions of known prototypes, and those versions could not have conquered the world because they were too unreliable, too costly, and, above all, too inefficient.

These inventions have succeeded because subsequent generations of engineers and scientists made them more efficient by requiring less energy for their operation, more durable (and less heavy) by using better (and lighter) materials in their construction, more reliable by devising higher-quality procedures and controls and, in modern electronics, by cramming more components on silicon microchips. Only continued technical innovation—the process that adopts, improves, and perfects inventions to make them widely applicable and easily affordable—made their mass-scale adoption (based on efficient and profitable operation) possible. I will illustrate this reality by concentrating on a single critical attribute of devices designed to convert energy: their efficiency, simply defined as the ratio of useful output per required input. Before I describe the impressive efficiency gains in electric lighting, I will trace the improvements of two widely deployed prime movers—steam engines, the energy converters that dominated the 19th century, and internal combustion engines, which are now installed in about 1.5 billion road vehicles and in tens of millions of other machines (from small lawnmowers to massive offroad trucks and ships).

Watt's improved steam engine was superior to Newcomen's older prototype, but its higher efficiency (2 percent vs. 0.5 percent for its predecessor), its large mass, and its operation under low pressure made it unsuitable for any mobile application: it could not power either trains or ocean vessels. That changed rapidly once the expiry of Watt's patent in 1800 led to the introduction of lighter and more efficient high-pressure compound engines, with two or three cylinders of increasing size to accommodate greater steam expansion and raise the efficiency. By 1850 the best compound steam engines in ocean shipping were more than 10 percent efficient, and in the early 1900s the best machines converted slightly more than 20 percent of coal's energy into motion.

The early gasoline internal combustion engines of the 1880s—used to power the first road vehicles built in Germany by Benz, Daimler, and Maybach—had compression ratios of just 2:1 to 3:1 and hence low thermal efficiencies of as little as 5 percent. A lifetime later, after the Second World War—when the wave of mass-scale car ownership began to spill from North America, first to Europe and then to Asia—internal combustion engines were about 25 percent efficient, and in 2025 Mazda's

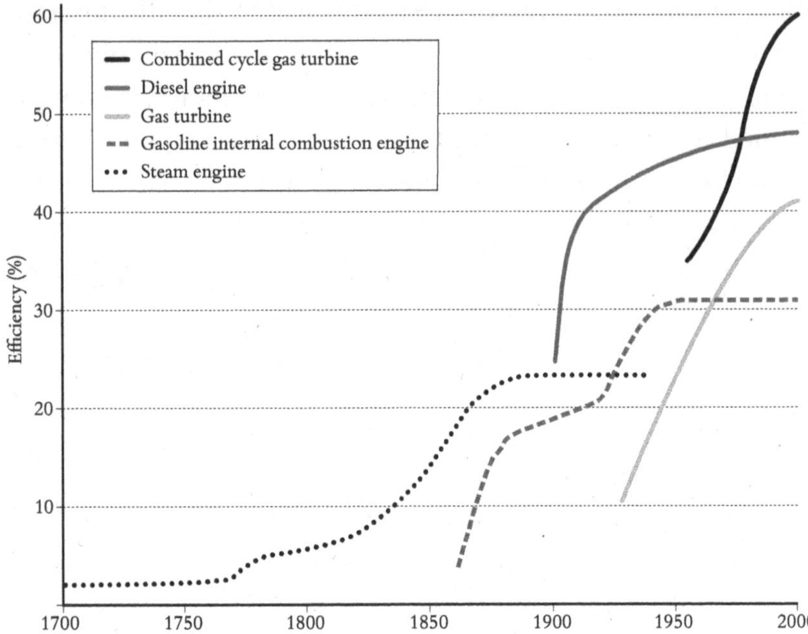

Modern prime movers, 1700–2000: a history of efficiency

Skyactiv-G design, using high compression ratios (up to 14:1, compared to the standard 8:1 to 10:1) that improve combustion and boost efficiency, converts just over 40 percent of fuel into engine output for an overall eightfold gain in fewer than 150 years. And while in 1897 the tests of Diesel's first engine recorded a thermal efficiency of 26 percent, today's best diesels powering large tankers and container ships have efficiencies of 50–52 percent, a doubling of performance in under 130 years.

Progress in the efficiency of lighting has been much more remarkable. Its best measure is luminous efficacy, expressed in lumens (lm, units of emitted light) per watt (W) of electricity. In 1879, Edison's first carbonized bamboo filament had an efficacy of a mere 1.4 lm/W; the first metallic (tungsten) filaments, which operated at higher temperature, were introduced in 1907 and raised that to 6–10 lm/W. By 1950 the best incandescent light bulbs had efficacies up to 17 lm/W, but by that time fluorescent lights, first introduced during the 1930s, reached more than 50 lm/W, and by the 1970s compact metal halide lights (producing an electric arc through a mixture of vaporized mercury and sodium iodide) raised the maximum to as much as 125 lm/W. Light-emitting diodes, the latest (and long-lasting) lighting innovation—using materials that emit light when an electric current passes through them—are an even better source of light, with efficacies up to 200 lm/W. This efficiency progress amounts to a nearly 150-fold improvement since 1879.

The implications of these developments are obvious. If energy conversion efficiencies remained at, or just slightly above, the level achieved by the pioneering devices, none of them could have experienced eventually mass-scale and profitable worldwide adoptions. Watt's engine could never have ushered in the first era of economic globalization (1860s–1914) because the ships would not have enough fuel for transatlantic journeys even if they carried mostly fuel and a small amount of cargo. We could never have reached the current level of automobilization if road vehicles were to require 5–8 times the amount of gasoline as they do today (to say nothing about the ensuing levels of air pollution). And Edison's original light bulbs could not have provided universal illumination because they would require 10 times more electricity than did the later metal filament bulbs and 100 times more electricity than today's best light sources.

And the same is true about historic mass reductions, be it the mass of the stationary internal combustion machines of jet engines (weight per unit of power, g/W) or, most spectacularly, for computers, with the reductions measured in grams per unit of memory (gram per kilobyte, g/kB). In 1981 IBM's first personal computer (12.7 kg, 64 kB) needed 198 g/kB. This book was written on a Dell Inspiron whose 8 gigabytes of RAM, mass of 1.87 kilograms, and 528 gigabytes of storage prorate to a mere 0.00000354 g/kB, or a reduction of seven orders of magnitude (about 56 million times!) in just over 40 years—another impressive illustration of the exceptional performance gains of modern solid-state electronics.

The conclusion is obvious: inventions are necessary—but they truly change the world only after decades of development have sufficiently improved their efficiency, enhanced their performance, or reduced their weight to levels that make universal adoption appealing, profitable, and inevitable. Eureka.

Europe's Irreversible Decline

History is full of reminders that a large population alone is no guarantee of economic might (think of the centuries of a decaying Ottoman Empire, or of the 27 years of China under Mao Zedong)—but it is impossible to be a globally important economic power and to have a prolonged influence on the world's technical, scientific, and artistic advances without it. The post–Second World War drive toward European integration had political, economic, and security reasons but it was also a clear recognition of this demographic reality. That is why population size and economic output are perhaps the most obvious starting combination for assessing the importance that countries or regions play in history. By those measures, Europe (in this assessment I am adopting a cultural rather than physical delimitation, by excluding the successor states of the USSR—that is, not counting Russia, Belarus, and Ukraine) had a glorious past (above all thanks to the amazing 19th century), is experiencing a troubled present (a downward trajectory is rarely pleasing, especially when it covers various measures) and its prospect is for a continuing decline across population, economic output, global clout, and cultural appeal.

In 1800, as advances in agriculture were putting an end to a long history of the continent's famines and as industrialization and new technical discoveries were beginning to transform traditional economies, Europe housed about 20 percent of the world's population and it generated a third (30–35 percent) of the global economic product. A century later, rapid advances in healthcare and hygiene and better diets and housing led to an unprecedented combination of declining infant mortalities and rising life expectancies, while the rapid progress of industrialization and urbanization (combined with the expansion of European empires) made the continent more populous (despite those massive emigration waves sent to the Americas and Australia) and economically much stronger. In 1900, with 25 percent of the global population, Europe produced 40–45 percent of the world's economic output. This European apogee ended abruptly in August 1914 as the self-inflicted madness of what became perhaps the most unnecessary (yet the most consequential) war in history—as it led to an even more devastating conflict just two decades later—began to engulf the continent.

Despite Europe's rapid post–Second World War recovery, the population growth surge in Asia, Africa, and Latin America (driven by high fertilities and improving life expectancies) reduced Europe's share to only about 12 percent of the global population by the beginning of the 21st century. But in the year 2000, Europe's intellectual and material contributions were still disproportionately important, producing no less than 20 percent of global economic output. In contrast, even the most favorable forecasts for 2050 indicate a further inexorable decline of the European population (driven by aging and declining rates of marriage, and by fertility shrinking to unprecedented levels) to no more than 7 percent of the global total, while the continent's share of the world's economic product will keep on receding, perhaps to as little as 10 percent.

As with any complex and large-scale phenomenon unfolding across decades, there are notable national and regional differences. The UK, the birthplace of the industrial revolution, has already become the most deindustrialized economy among Europe's major nations: in 2025, less than 9 percent of its GDP came from manufacturing compared to 16 percent in Italy and 19 percent in Germany. And at about 1.8 children per family, France's total fertility rate is still close to the replacement level of 2.1, while the fertilities now prevailing in Spain (1.16) or Latvia (1.27) have sunk so low that there is no realistic prospect for their eventual recovery. Moreover, in many depopulating regions across Europe the fertilities have declined below 1.0 to lows never seen before in the peacetime history of humanity.

Europe's population retreat shows a complex spatial pattern, with the most depopulated regions including large parts of the Romanian and Bulgarian countryside in the continent's southeast, in Spain's interior and much of Portugal in the Atlantic west, as well as throughout the Italian Mezzogiorno, in all Baltic states, in eastern Finland, in most of rural Poland, and starkly, in the former East Germany (Berlin and its immediate environs being the only exception). The Spanish countryside has lost 30 percent of its population since Franco's death in 1975, with the greatest declines in the country's interior including the provinces of Castilla y León, Galicia, Aragon, Extremadura, and parts of Castilla–La Mancha. These regions now constitute a new *España vacíada* of nearly empty villages and small towns, abandoned farmhouses, overgrown orchards, and decaying factories—many within just an hour's drive of Madrid.

And depopulation is accompanied by continent-wide aging. By 2050, four out of six countries with the highest share of the population above the retirement age of 65 will be in Europe: South Korea will lead with 41 percent, followed by

An abandoned village in Italy

38 percent in Japan, 37 percent in Italy and Spain, 35 percent in Greece, and 34 percent in Portugal. This inexorable aging shift will deepen well-known economic and social complications ranging from securing adequate pensions with a shrinking taxation base to staffing increasingly overloaded healthcare services. Policies favoring legal mass-scale immigration could slow down and then reverse the depopulation and aging, but even if they were widely welcome (not the case anywhere in Europe), the influx of migrants from Africa, the Middle East, and Southeast Asia might do little either for deindustrialization or for rural depopulation as immigrants are attracted overwhelmingly by large cities.

What a trajectory it has been, and what lies ahead! In 1900 Europe was the most enviable continent, whose quarter of the world's population produced nearly half the economic output and accounted for even higher shares of that incomparable late 19th-century wave of scientific and technical progress, and for admirable achievements in the arts—ranging from grand novels to amazing paintings. Europe produced a brilliant civilization whose achievements have influenced the entire world—but by 2025 the continent had changed into a declining and deindustrializing contributor to the global economy and the museum of the world (subject to seasonally extreme waves of overtourism) whose shrinking population (its total is now the equivalent of Indonesia and Nigeria) and ebbing economic fortunes signal further retreats and additional complications by the mid-21st century. *Sic transit gloria . . .*

Where We Stand

We have never had so much—and so readily and almost freely available—information as we have during the third decade of the 21st century. At the same time, we have never had entire populations so poorly informed, indeed so misinformed, about the fundamental realities of our world. That contrast is easily explained: on the one hand there is a litany of statistics showing steady, and substantial, declines in book reading, above all among young people, while those adults and youngsters who still read do so less often. On the other hand there are the soaring numbers showing the daily frequency of checking a mobile phone (in the US, now about 200 times a day) and engaging in snippet-reading on screens or, more often, watching inane images—with much of the "content" supplied by self-appointed instant experts, influencers, and clueless commentators posting without any checks, references, or reviews.

This book is yet another (quixotic?) attempt to change the trend, to bring not only more of the real world into a daily information stream but to do so by doggedly quantifying the realities, by relying on the best available statistics, measurements, and studies to explain the world through numbers. And what would be the ideal outcome, now that you have read this book? If, trying to understand anything new or old, you were to ask not only how and why but also how much, how far, how fast, how old, how good, how bad, how costly, or how helpful. Simply: if you were to seek, on your own, the numbers behind the stories.

Further Reading

A MARRIAGE OF MISCONCEPTIONS AND REALITY

Deswal, Vageshwari, "Prohibition of Child Marriage (Amendment) Bill, 2021: A critique," *Times of India* (January 25, 2022), https://timesofindia.indiatimes.com/blogs/legally-speaking/prohibition-of-child-marriage-amendment-bill-2021-a-critique/.

Reid, Alice, "What age did people marry in the British past?", Cambridge Group for the History of Population and Social Structure, University of Cambridge, https://www.campop.geog.cam.ac.uk/blog/2024/07/11/what-age-did-people-marry/.

UNICEF, "Child Marriage—UNICEF DATA," https://data.unicef.org/topic/child-protection/child-marriage/.

Xu, Kim Qinzi, "Changing patterns and determinants of first marriage over the history of the People's Republic of China," *Population* 74 (2019), pp. 207–236.

FERTILITY IN RETREAT

Population Reference Bureau, "Total fertility rate" (updated 9/11/2024), https://www.prb.org/international/indicator/fertility/map/country.

Skirbekk, Vergard, *Decline and Prosper!: Changing Global Birth Rates and the Advantages of Fewer Children* (London: Palgrave Macmillan, 2022).

United Nations, "World Population Prospects 2024," https://population.un.org/wpp/.

Winter, Jay, and Michael Teitelbaum, *The Global Spread of Fertility*

Decline: Population, Fear, and Uncertainty (New Haven, CT: Yale University Press, 2013).

LIVING FOREVER

Dong, Xiao, Brandon Milholland, and Jan Vijg, "Evidence for a limit to human lifespan," *Nature* 538/7624 (2013), pp. 257–259, https:doi.org/10.1038/nature19793.

Kurzweil, Ray, and Terry Grossman, *Fantastic Voyage: Live Long Enough to Live Forever* (Emmaus, PA: Rodale Books, 2004).

Olshansky, S. Jay, Bradley J. Willcox, Lloyd Demetrius, and Hiram Beltrán-Sánchez, "Implausibility of radical life extension in humans in the twenty-first century," *Nature Aging* 4 (2024), pp. 1635–1642, https://doi.org/10.1038/s43587-024-00702-3.

Nippon.com, "More Japanese households with seniors caring for seniors" (July 27, 2023), https://www.nippon.com/en/japan-data/h01733/.

Statistics Canada, "Life expectancy and other elements of the complete life table, three-year estimates, Canada, all provinces except Prince Edward Island" (December 4, 2024), https://www150.statcan.gc.ca/t1/tbl1/en/tv.action?pid=1310011401/.

Wilmoth, John R., "How long can we live? A review essay," *Population and Development Review* 27/4 (2001), pp. 791–800, http://www.jstor.org/stable/2695190.

TOO MUCH OF A GOOD THING

Dattani, Saloni, "What was the Golden Age of Antibiotics, and how can we spark a new one?", Our World in Data (December 22, 2024), https://ourworldindata.org/golden-age-antibiotics.

Nicolaou, Kyriacos C., and Stephan Rigol, "A brief history of antibiotics and select advances in their synthesis," *The Journal of Antibiotics* 71 (2018), pp. 153–184, https://doi.org/10.1038/ja.2017.62.

Minasyan, Hayk, "Sepsis: mechanisms of bacterial injury to

the patient," *Scandinavian Journal of Trauma, Resuscitation and Emergency Medicine* 27/1 (2019), p. 19, https://doi.org/10.1186/s13049-019-0596-4.

Mora, Camilo, et al., "Over half of known human pathogenic diseases can be aggravated by climate change," *Nature Climate Change* 12 (2022), pp. 869–875, https://doi.org/10.1038/s41558-022-01426-1.

Shi, Zhenfeng, et al., "Comprehensive overview of the antibiotics approved in the last two decades: Retrospects and prospects," *Molecules* 28/4 (2023), p. 1762, https://doi.org/10.3390/molecules28041762.

World Health Organization, "1 in 3 use antibiotics without prescription, WHO/Europe's study shows" (November 21, 2022), https://www.who.int/europe/news-room/21-11-2022-1-in-3-use-antibiotics-without-prescription–who-europe-s-study-shows.

HOW BAD WAS COVID-19?

Mostert, Saskia, Marcel Hoogland, Minke Huibers, and Gertjan Kaspers, "Excess mortality across countries in the Western World since the COVID-19 pandemic: 'Our World in Data' estimates of January 2020 to December 2022," *BMJ Public Health*, 2 (2024), e000282, https://doi.org/10.1136/bmjph-2023-000282.

O'Driscoll, Megan, et al., "Age-specific mortality and immunity patterns of SARS-CoV-2," *Nature* 590 (2021), pp. 140–145, https://doi.org/10.1038/s41586-020-2918-0.

Mathieu, Edouard, "Excess mortality during the Coronavirus pandemic (COVID-19)," Our World in Data (2020, revised in 2024), https://ourworldindata.org/excess-mortality-covid.

Smil, Vaclav, *Global Catastrophes and Trends* (Cambridge, MA: MIT Press, 2009).

Wang, Haidong, et al., "Estimating excess mortality due to the COVID-19 pandemic: a systematic analysis of COVID-19-related mortality, 2020–21," *The Lancet* 399/10334 (2022), pp. 1513–1536, https://www.thelancet.com/article/S0140-6736(21)02796-3/fulltext.

CAN WE MEASURE HAPPINESS?

Davies, William, *The Happiness Industry: How the Government and Big Business Sold Us Well-Being* (New York: Verso Books, 2015).

Helliwell, John F., et al. (eds), *World Happiness Report 2024* (Oxford: University of Oxford Wellbeing Research Centre, 2024), https://www.worldhappiness.report/ed/2024/.

Jardine, Boris, "The origins of happiness," *The Total Archive* 6 (2016), https://limn.press/article/the-origins-of-happiness.

Stewart, Frances, "Against happiness: A critical appraisal of the use of measures of happiness for valuating progress in development," *Journal of Human Development and Capabilities* 15/4 (2014), pp. 293–307, https://doi.org/10.1080/19452829.2014.903234.

ARE WE ON THE WAY TO A VEGAN PLANET?

Food and Agriculture Organization of the United Nations, *Meat Market Review* (Rome: FAO, 2021), https://openknowledge.fao.org/server/api/core/bitstreams/2a4c3fe0-0c7d-4418-981b-2471d4924fef/content.

Food and Agriculture Organization of the United Nations, FAOSTAT, https://www.fao.org/faostat/en/#data (accessed 2025).

Kirchner, Jody, Daniel Gertner, and Emma Ignaszewski, "U.S. retail market insights for the plant-based industry," Good Food Institute (2024), https://gfi.org/marketresearch/.

Minassian, Liana, "Why the global rise in vegan and plant-based eating is no fad (30x increase in US vegans + other astounding vegan stats)," Food Revolution Network (April 6, 2022), https://foodrevolution.org/blog/vegan-statistics-global/.

Ritchie, Hannah, Pablo Rosado, and Max Roser, "Meat and dairy production," Our World in Data (August 2017, updated December 2023), https://ourworldindata.org/meat-production.

FEEDING RUSSIA

Dent, David, Boris Boincean, Yuriy Dmytruk, and Zhanguo Bai, "A candle in the wind," *International Journal of Environmental Studies* 79/4 (2022), pp. 587–596, https://doi.org/10.1080/00207233.2022.2085454.

Goodwin, Barry, and Thomas Grennes, "Tsarist Russia and the world wheat market," *Explorations in Economic History* 35/4 (1998), pp. 405–430.

US Department of Agriculture, Foreign Agricultural Service, "Russia Wheat: Record MY 2022/23 Harvest," *Commodity Intelligence Report* (April 11, 2023), https://ipad.fas.usda.gov/highlights/2023/04/Russia/index.pdf.

Welsh, Caitlin, and Joseph Glauber, *Food as the "Silent Weapon" Russia's Gains and Ukraine's Losses* (Washington, DC: Center for Strategic and International Studies, 2024), https://csis-website-prod.s3.amazonaws.com/s3fs-public/2024-02/240229_Welsh_Silent_Weapon_0.pdf?VersionId=Ss6dcxYB_dIk7i80O9xkZhVGpdvP_h8R.

WHY JAPANESE MILK IS MORE IMPORTANT THAN ITS FISH

Carletti, Francesco, *Ragionamenti di Francesco Carletti Fiorentino sopra le cose da lui vedute ne' suoi viaggi dell' Indie Orientali, e d'altri paesi* (Florence: Nel Garbo, nella stamperia di G. Manni, 1701), https://archive.org/details/ragionamentioocarl.

Food and Agriculture Organization of the United Nations, FAOSTAT, https://www.fao.org/faostat/en/#data (accessed 2025).

Japan Dairy Association, *The History of Milk in Japan* (Tokyo: J-Milk, 2020), https://www.j-milk.jp/report/study/h40gb4000000011y2-att/h40gb40000003f7e.pdf.

Smil, Vaclav, and Kazuhiko Kobayashi, *Japan's Dietary Transition* (Cambridge, MA: MIT Press, 2012).

Storhaug, Christian Løvold, Svein Kjetil Fosse, and Lars T. Fadnes, "Country, regional, and global estimates for lactose malabsorption in adults: a systematic review and meta-analysis," *Lancet Gastroenterology and Hepatology* 2/10 (2017), pp. 738–746, https://doi.org/10.1016/S2468-1253(17)30154-1.

THE MADNESS OF BLUEBERRIES FROM PERU

Blueberries Consulting, "The blueberry growth route in Peru," *Blue Magazine* (May 13, 2005), https://blueberriesconsulting.com/en/la-ruta-de-crecimiento-del-arandano-en-el-peru/.

Coville, Frederick V., *Experiments in Blueberry Culture* (Washington, DC: USDA Bulletin, vol. 193, 1910).

Environmental Working Group, "EWG Shopper's Guide: Dirty Dozen," EWG Shopper's Guide to Pesticides in Produce (2025), https://www.ewg.org/foodnews/dirty-dozen.php.

Macha Huamán, Roberto, Fabiola Cruz Navarro Soto, Alejandro Ramírez Ríos, and Emigdio Antonio Alfaro Paredes, "International market concentration of fresh blueberries in the period 2001–2020," *Humanities and Social Sciences Communications* 10/1 (2023), https://doi.org/10.1057/s41599-023-02455-7.

US Department of Agriculture, Foreign Agricultural Service, "Blueberries Around the Globe—Past, Present, and Future," USDA.gov (October 21, 2021), https://www.fas.usda.gov/data/blueberries-around-globe-past-present-and-future.

THE UNIVERSAL TREND IN THE COST OF FOOD

Chen, Menggen, "Engel's law in China: Some new evidence," *Review of Development Economics* 26/3 (2022), pp.1640–1662, https://doi.org/10.1111/rode.12887.

Engel, Ernst, "Die Productions- und Consumtionsverhältnisse des Königreichs Sachsen," *Zeitschrift des statistischen Bureaus des Königlich Sächsischen Ministerium des Inneren* 8 (1857), pp. 1–54.

Houthakker, Hendrik S., "An international comparison of household expenditure patterns, commemorating the centenary of Engel's Law," *Econometrica* 25/4 (1957), pp. 532–551, https://www.jstor.org/stable/1905382.

US Department of Agriculture, Food Expenditure Series (updated December 19, 2025), https://www.ers.usda.gov/data-products/food-expenditure-series.

US Department of Labor, *100 Years of U.S. Consumer Spending Data for the Nation, New York City, and Boston* (Washington, DC: US Department of Labor, 2006), https://www.bls.gov/opub/100-years-of-u-s-consumer-spending.pdf.

THE ARCHITECTURE OF AIR POLLUTION

A Hauf Stop, "When Glasgow had the Tallest Chimney(s) in the World," A Hauf Stop [blog] (July 20, 2021), https://ahaufstop.blogspot.com/2021/07/when-glasgow-had-tallest-chimneys-in.html.

Greenwood, Cedric, *Dockland, Smokestacks and Slums: In the Shadows of British Industry* (Horncastle, Lincolnshire: Mortons Books, 2024).

Wikipedia, "List of tallest chimneys," https://en.wikipedia.org/wiki/List_of_tallest_chimneys.

FUELS VS. BATTERIES

Automotive Manufacturing Solutions, "Powering the future: Top 5 EV battery chemistries and formats across the world" (July 24, 2024), https://www.automotivemanufacturingsolutions.com/electrification/powering-the-future-top-5-ev-battery-chemistries-and-formats-across-the-world/542508.

Batemo, "Panasonic NCR2170-M," https://www.batemo.com/products/batemo-cell-explorer/panasonic-ncr2170-m/.

Blain, Loz, "Amprius ships first batch of 'world's highest density' batteries," New Atlas (February 13, 2022), https://newatlas.com/energy/amprius-450-wh-kg-battery/.

Bower, George, "Tesla Model 3 2170 energy density compared to Bolt, Model S P100D," InsideEVs (February 7, 2019), https://insideevs.com/news/342679/tesla-model-3-2170-energy-density-compared-to-bolt-model-s-p100d/.

Clean Energy Insitute, "Lithium-ion battery," University of Washington, https://www.cei.washington.edu/research/energy-storage/lithium-ion-battery/.

Panasonic Group, "Panasonic Energy begins mass production at new automotive lithium-ion battery factory in Kansas, aiming for annual capacity of 32 GWh to accelerate US local production," Panasonic Energy Co., Ltd. (July 14, 2025), https://news.panasonic.com/global/press/en250714-6.

Turner, James Morton, *Charged: A History of Batteries and Lessons for a Clean Energy Future* (Seattle, WA: University of Washington, 2022).

THE RISE OF LIQUIFIED NATURAL GAS

Esanda, "LNG History," https://www.esandaengineering.com/Useful-Information/lng-history.html.

Energy Institute, *Statistical Review of World Energy* (Energy Institute, 2025), https://www.energyinst.org/statistical-review.

International Gas Union, *2024 World LNG Report* (International Gas Union, 2024), https://www.igu.org/igu-reports/2024-world-lng-report.

QatarEnergy (formerly QatarGas), https://www.qatarenergylng.qa/english/.

THE GLOBAL ENERGY TRANSITION WILL PROFOUNDLY TRANSFORM MODERN CIVILIZATION (SLOWLY)

Amadeo, Kimberley, "US GDP by year, compared to recessions and events," The Balance (updated May 26, 2024), https://www.thebalancemoney.com/us-gdp-by-year-3305543.

CTBTO Preparatory Commission, https://www.ctbto.org.

Dreier, Casey, "An improved cost analysis of the Apollo Program," *Space Policy* 60 (2022), 101476, https://doi.org/10.1016/j.spacepol.2022.101476.

Fishback, Price, "World War II in America: Spending, deficits, multipliers, and sacrifice," Center for Economic Policy Research (November 12, 2019), https://cepr.org/voxeu/columns/world-war-ii-america-spending-deficits-multipliers-and-sacrifice.

International Copper Association, "Copper needs" (June 2017), https://internationalcopper.org/wp-content/uploads/2017/06/2017.06-E-Mobility-Factsheet-1.pdf.

Krishnas, Mekala, et al., "The economic transformation: What would change in the net-zero transition," McKinsey Sustainability (January 25, 2022), https://www.mckinsey.com/capabilities/sustainability/our-insights/the-economic-transformation-what-would-change-in-the-net-zero-transition.

Smil, Vaclav, *Energy Transitions: Global and National Perspectives* (Santa Barbara, CA: Praeger, 2017).

Smil, Vaclav, *Halfway Between Kyoto and 2050: Zero Carbon Is a Highly Unlikely Outcome* (Vancouver: Fraser Institute, 2024).

World Bank Group, "GDP (current US$)—United States," https://data.worldbank.org/indicator/NY.GDP.MKTP.CD?locations=US (accessed 2025).

MICROCHIPS HAVE GROWN RAPIDLY, BUT SOMETHING HAS EXPLODED EVEN FASTER

Dhari, Rahul, "TNT Equivalent Calculator," https://www.omnicalculator.com/physics/tnt-equivalent (accessed 2025).

Specktor, Brandon, "Russia declassifies footage of 'Tsar Bomba'—the most powerful nuclear bomb in history," Live Science (November 4, 2022), https://www.livescience.com/tsar-bomba-secret-test-footage-declassified.html.

Maienschein, Jon L., *Estimating Equivalency of Explosives Through a Thermochemical Approach* (Livermore, CA: Lawrence Livermore Laboratory, US Department of Energy, 2022), https://www.osti.gov/servlets/purl/15002352.

Malik, John, *The Yields of Hiroshima and Nagasaki Explosions* (Los Alamos, NM: Los Alamos National Laboratory, 1985), http://atomicarchive.com/Docs/pdfs/00313791.pdf.

US Arms Control and Disarmament Agency, "Note 1: Nuclear Weapons Yield," *Worldwide Effects of Nuclear War* (Washington, DC: US Arms Control and Disarmament Agency, 1975), https://www.atomicarchive.com/resources/documents/effects/wenw/note1.html.

WHY GREEN ROADS ARE ABOUT MORE THAN THE VEHICLES ON THEM

Anupam, Kumar, Daniel Akinmade, Cor Kasbergen, Sandra Erkens, and Festus Adebiyi, "A state-of-the-art review of Natural bitumen in pavement: Underlining challenges and the way forward," *Journal of Cleaner Production* 382 (2023), https://doi.org/10.1016/j.jclepro.2022.134957.

Burdon, Judson, "World's top asphalt producing countries: Pictorial overview," Asphalt Kingdom (September 11, 2025), https://blog.asphaltkingdom.com/top-asphalt-producers.

European Asphalt Pavement Association, *Asphalt In Figures* archive, https://eapa.org/asphalt-in-figures-archive/.

Observatory of Economic Complexity, "Bitumen and Asphalt," https://oec.world/en/profile/hs/bitumen-and-asphalt.

Pavement Interactive, "Asphalt production and oil refining," https://pavementinteractive.org/reference-desk/materials/asphalt/asphalt-production-and-oil-refining/.

Speight, James, "Newly laid asphalt paving and eventual deterioration," in *Asphalt Materials Science and Technology* (Amsterdam: Elsevier, 2015).

Sun, Lijun, *Structural Behavior of Asphalt Pavements: Integrated Analysis and Design of Conventional and Heavy Duty Asphalt Pavement* (Amsterdam: Elsevier, 2016).

Federal Highway Administration, "Asphalt Pavement Recycling with Reclaimed Asphalt Pavement (RAP)," US Department of Transportation (2020), https://www.fhwa.dot.gov/pavement/recycling/rap/.

Venditti, Bruno, "The road to decarbonization: How asphalt is affecting the planet," Visual Capitalist (May 13, 2022), https://www.visualcapitalist.com/sp/the-road-to-decarbonization-how-asphalt-is-affecting-the-planet/.

A GRAIN OF TRUTH

Kitazume, Masaki, "Sustainable land reclamation in coastal area," *Revue française de Géotechnique* 170/2 (2022), https://doi.org/10.1051/geotech/2021033.

Morita, Nobua, *Geomechanics of Sand Production and Sand Control* (Houston, TX: Gulf Professional Publishing, 2022).

Ser, Kuang Keng Kuek, Krisna Pradipta, Budhy Nurgianto, and Nabilah Said, "Reclamation: A flawed solution," Beneath the Sands (May 4, 2023), https://www.beneaththesands.earth/reclamation.

US Geological Survey, "Sand and Gravel (Construction)," Mineral

Commodity Summaries (January 2024), https://pubs.usgs.gov/periodicals/mcs2024/mcs2024-sand-gravel.pdf.

YOU MIGHT NOT HAVE THOUGHT MUCH ABOUT SILICON—BUT IT IS INDISPENSABLE

Institute of Supply Chain Management, "Industrial application of silicone: An overview" (October 26, 2022), https://www.ioscm.com/blog/industrial-application-of-silicone-an-overview/.

Pastore, Guido, et al., "Provenance and recycling of Sahara Desert sand," *Earth-Science Reviews* 216 (2021), 103606, https://doi.org/10.1016/j.earscirev.2021.103606.

University Wafer, "Czochralski & float zone silicon ingots to manufacture wafers," https://www.universitywafer.com/silicon-ingots.html.

FROM GLASS BEADS TO MASSIVE WINDOWS

Bricknell, David, *Float: Pilkington's Glass Revolution* (Lancaster: Crucible Books, 2009).

Pilkington, "Invention of Float Glass," https://www.pilkington.com/en-gb/uk/about/heritage/invention-of-float-glass#.

Qingdao Morn Building Materials, "Guinness world record–world largest glass window," https://www.mornglass.com/guinness-world-record-world-largest-glass-window.html.

National Glass Association, World of Glass Map, 2025, glass.org/world-glass-map

Uusitalo, Olavi, *Float Glass Innovation in the Flat Glass* (Amsterdam: Elsevier, 2014).

ENDLESS STREAMS

Benyathiar, Patnarin, et al., "Polyethylene terephthalate (PET) bottle-to-bottle recycling for the beverage industry: A review," *Polymers* 14/12 (2022), 2366, https://doi.org/10.3390/polym14122366.

Joseph, Tomy Muringayil, et al., "Polyethylene terephthalate (PET) recycling: A review," *Case Studies in Chemical and Environmental Engineering* 9 (2024), 100673, https://doi.org/10.1016/j.cscee.2024.100673.

Thomas, Sabu, et al. (eds), *Recycling of Polyethylene Terephthalate Bottles* (Amsterdam: Elsevier, 2018).

WHY E-WASTE IS WORTH MORE THAN ITS WEIGHT IN GOLD ORE

MEC Mining, "Top 10 metals and minerals powering your mobile phone" (April 17, 2024), https://www.mecmining.com.au/the-10-essential-metals-and-minerals-behind-your-mobile-phone/.

Sahan, Merve, et al., "Determination of metal content of waste mobile phones and estimation of their recovery potential in Turkey," *International Journal of Environmental Research and Public Health* 16 (2019), p. 887, https://doi.org/10.3390/ijerph16050887.

United Nations Institute for Training and Research, *The Global E-waste Monitor 2024* (Geneva: UNITAR, 2024), https://ewastemonitor.info/the-global-e-waste-monitor-2024/.

World Health Organization, "Electronic waste (e-waste)" (October 1, 2024), https://www.who.int/news-room/fact-sheets/detail/electronic-waste-(e-waste).

BIGGER AND BIGGER CARS

Ford, "2025 F-150," https://www.ford.com/trucks/f150/2025/?gnav=header-trucks-vhp.

Model T Ford Club of America, https://www.mtfca.com.

Smil, Vaclav, *Creating and Transforming the 20th Century* (New York: Oxford University Press, 2025).

US Department of Energy, "Vehicle Weight Classes & Categories," Alternative Fuels Data Center (2025), https://afdc.energy.gov/data/10380.

US Environmental Protection Agency, *The 2024 EPA Automotive Trends Report* (Washington, DC: EPA, 2024), https://nepis.epa.gov/Exe/ZyPDF.cgi?Dockey=P101CUU6.pdf.

AN UNDERAPPRECIATED TECHNICAL REVOLUTION

Kendall, Lane C., and James J. Buckley, *The Business of Shipping* (Centreville, MD: Cornell Maritime Press, 2001).

Levinson, Marc, *The Box: How the Shipping Container Made the World Smaller and the World Economy Bigger* (Princeton, NJ: Princeton University Press, 2016).

Marine Traffic, "MSC *Irina*," https://www.marinetraffic.com/en/ais/details/ships/shipid:7507138/mmsi:636022601/imo:9929429/vessel:MSC_IRINA (accessed 2025).

FLYING BETTER, NOT FASTER

Kupietzky, Joshua, "ETOPS: History, Evolution, Current Applications," *Airways* (April 11, 2023), https://www.airwaysmag.com/legacy-posts/etops-history-evolution-applications.

Smil, Vaclav, *Creating and Transforming the 20th Century* (New York: Oxford University Press, 2025).

Tennekes, Henk, *The Simple Science of Flight* (Cambridge, MA: MIT Press, 2009).

US Department of Transportation, "Extended Operations (ETOPS) of Multi- Engine Airplanes," *Federal Register* 72/9 (2007), https://www.govinfo.gov/content/pkg/FR-2007-01-16/pdf/07-39.pdf.

THE LONG HISTORY AND UNCERTAIN FUTURE OF ELECTRIC VEHICLES

McKerracher, Colin, et al., *Electric Vehicle Outlook 2025* (Bloomberg NEF, 2025), https://about.bnef.com/insights/clean-transport/electric-vehicle-outlook/#overview.

International Energy Agency, *Global EV Outlook 2024* (Paris: IEA, 2024), https://www.iea.org/reports/global-ev-outlook-2024.

Kirsch, David A., *The Electric Vehicle and the Burden of History* (New Brunswick, NJ: Rutgers University Press, 2000).

Nolan, Stella, "Copper shortage could derail global 2050 EV goals," *EV Magazine* (August 28, 2024), https://evmagazine.com/sustainability/copper-shortage-could-derail-global-2050-ev-goals.

Rissman, Jeffrey, "The future of electric vehicles in the US to 2050," Energy Innovation (September 13, 2017), https://energyinnovation.org/report/future-electric-vehicles-u-s-2050/.

WHY IT'S NOT JUST FOREIGN WAGES THAT MAKE FARAWAY-MADE PRODUCTS CHEAPER

American Transportation Research Institute, *An Analysis of the Operational Costs of Trucking: 2024 Update* (ATRI, 2024), https://truckingresearch.org/2024/06/an-analysis-of-the-operational-costs-of-trucking-2024-update/.

Condon, Patrick M., and Karl Dow, "A cost comparison of transportation modes," *Foundational Research Bulletin* (2009), https://www.researchgate.net/publication/237126920_A_Cost_Comparison_of_Transportation_Modes.

RSI Logistics, "Comparing the Costs of Rail Shipping vs Truck" (February 6, 2024), https://www.rsilogistics.com/blog/comparing-the-costs-of-rail-shipping-vs-truck/.

Wolf, Devorah, "Free Container Shipping Cost Calculator," Freightos (September 4, 2025), https://www.freightos.com/freight-resources/container-shipping-cost-calculator-free-tool/.

THE FIRST AGE OF GLOBALIZATION

Faubert, Violaine, *Learning from the First Globalisation (1870–1914)* (Paris: Ministère de l'Économie, des Finances et de l'Industrie Direction Générale du Trésor, 2011).

Estevadeordal, Antoni, Brian Frantz, and Alan M. Taylor, "The rise and fall of world trade, 1870–1939," *The Quarterly Journal of Economics* 118/2 (2003), pp. 359–407, https://doi.org/10.1162/003355303321675419.

Fouquin, Michel, and Jules Hugot, *Back to the Future: International Trade Costs and the Two Globalizations*, CEPII Working Paper (2016), https://www.cepii.fr/PDF_PUB/wp/2016/wp2016-13.pdf.

Keynes, John Maynard, *Economic Consequences of Peace* (London: Macmillan, 1920).

O'Rourke, Kevin H., and Jeffrey Williamson, "When did globalisation begin?", *European Review of Economic History* 6/1 (2002), pp. 23–50.

COSTLIEST EXPORTS: FROM DIAMONDS TO GLASS EELS

Auguste Escoffier School of Culinary Arts, "World's most expensive foods: 10 of the priciest ingredients on the planet" (December 12, 2024), https://www.escoffier.edu/blog/world-food-drink/worlds-most-expensive-foods/.

Dubai Sotheby's International Realty, "The Most Expensive Diamonds in the World," The Journal, https://sothebysrealty.ae/the-journal/most-expensive-diamond-in-the-world/.

Food and Agriculture Organization of the United Nations, "Crops and Livestock Products, " https://www.fao.org/faostat/en/#data/TCL (accessed 2025).

Neto, Veronica, "Giant bluefin tuna sells for $3.1 million in Tokyo," *Fortune* (January 6, 2019), https://fortune.com/2019/01/06/bluefin-tuna-japan-auction/.

Observatory of Economic Complexity, "Precious stones," https://oec.world/en/profile/hs/precious-stones.

WHY DO SOME COUNTRIES GET MORE INVESTMENT THAN OTHERS?

Collins, David, *An Introduction to International Investment Law* (Cambridge: Cambridge University Press, 2023).

Graham, Edward M., *Foreign Direct Investment in the World Economy* (Washington, DC: International Monetary Fund, 2016).

Morrison, Wayne, *China's Economic Rise: History, Trends, Challenges, and Implications for the United States* (Washington, DC: Congressional Research Service, 2019).

Hanemann, Thilo, Daniel H. Rosen, and Cassie Gao, *Two-Way Street: 25 Years of US-China Direct Investment* (Rhodium Group, 2016).

THE MOST IMPORTANT VARIABLE FOR UNDERSTANDING STALLING GROWTH

Congressional Budget Office, "CBO Projection: Total Factor Productivity Growth," https://www.ceicdata.com/en/united-states/real-total-factor-productivity-projection-congressional-budget-office/cbo-projection-total-factor-productivity-growth (accessed 2025).

Federal Reserve Bank of St. Louis, "Manufacturing Sector: Labor Productivity (Output per Hour) for All Workers," https://fred.stlouisfed.org/series/OPHMFG (accessed 2025).

Gordon, Robert J., "Does the 'New Economy' measure up to the great inventions of the past?", *Journal of Economic Perspectives* 14 (2000), pp. 49–74.

Gordon, Robert J., *The Rise and Fall of American Growth*, (Princeton, NJ: Princeton University Press, 2016).

Shackleton, Robert, *Total Factor Productivity Growth in Historical Perspective* (Washington, DC: Congressional Budget Office, 2013).

Solow, Robert M., "Technical change and the aggregate production," *Review of Economics and Statistics* 39/3 (1957), pp. 313–320.

Solow, Robert M., "Growth theory and after," Prize Lecture (The Nobel Foundation, 1987), http://nobelprize.org/nobel_prizes/economics/laureates/1987/solow-lecture.html.

Solow, Robert M., "We'd better watch out," *New York Times Book Review* (July 12, 1987), p. 36.

WHY THE BLACK ECONOMY IS MUCH BIGGER THAN YOU MIGHT THINK

Medina, Leandro, and Freidrich Schneider, *Shedding Light on the Shadow Economy: A Global Database and the Interaction with the Official One* (Munich: Munich Society for the Promotion of Economic Research—CESifo GmbH, 2019).

Marshall, Emily C., James Saunoris, Mario Solis-Garcia, and Trang Do, "Measuring the size and dynamics of U.S. state-level shadow economies using a dynamic general equilibrium model with trends," *Journal of Macroeconomics* 75 (2023), 103491, https://doi.org/10.1016/j.jmacro.2022.103491.

Schneider, Friedrich, *Size, Causes and Consequences of the Underground Economy: An International Perspective* (London: Routledge, 2005).

WHAT USE IS COMPLEXITY?

Hausmann, Ricardo, et al., *The Atlas of Economic Complexity: Mapping Paths to Prosperity*, (Cambridge, MA: Puritan Press, 2011).

Hidalgo, Cesar A., and Ricardo Hausmann, "The Building Blocks of Economic Complexity," *Proceedings of the National Academy of Sciences* 106/26 (2009), pp. 10570–10575, https://doi.org/10.1073/pnas.0900943106.

Observatory of Economic Complexity, https://oec.world/en (accessed 2025).

THE RISE AND FALL OF PORTUGAL

Camões, Luís Vaz de, *Os Lusíadas* (1572). Notable English translators include J. J. Aubertin, Richard Burton, William C. Atkinson, Guy Bulter, and Landeg Whie.

Couto, Jorge (ed.), *Historical Atlas of Portugal: A Geopolitical Perspective* (Lisbon: Sociedade Francisco Manuel dos Santos, 2022).

Crowley, Roger, *Conquerors: How Portugal Forged the First Global Empire* (New York: Random House, 2015).

Disney, Anthony R., *A History of Portugal and the Portuguese Empire: From Beginnings to 1807* (Cambridge: Cambridge University Press, 2009).

Slave Voyages, https://www.slavevoyages.org/voyage/all-voyages#voyages (accessed 2025).

A HISTORY OF CRANES (MACHINES, NOT BIRDS)

Krøll Cranes, *K10000*, https://www.krollcranes.dk/media/k-10000.pdf.

Liebherr Group, https://www.liebherr.com/.

Marks, Edward C. R., *Construction of Cranes and Other Lifting Machinery* (Manchester: Technical Publishing Company, 1904).

Musei Vaticani, "Mausoleum of the Haterii," https://www.museivaticani.va/content/museivaticani/en/collezioni/musei/museo-gregoriano-profano/Mausoleo-degli-Haterii.html.

THE PRACTICAL LIMITS OF DESIGN

Bishop, Chris, *Ships: The History and Specifications of 300 World-Famous Ships* (London: Amber Books, 2020).

M.B. Marsh Marine Design, "Length-beam ratio," https://marine.marsh-design.com/content/length-beam-ratio.

McKay, Richard C., *Some Famous Sailing Ships and their Builder Donald McKay* (Caledonian Maritime Research Trust, 1928).

EXTREME DESIGNS

Ahmed, Zahra, "5 Biggest Oil Tankers Which Are Now |Scrapped," Marine Insight (February 25, 2022), https://www.marineinsight.com/maritime-history/5-biggest-oil-tankers-which-are-now-scrapped/.

Evergreen Aviation & Space Museum, "The Spruce Goose Lives On!", https://www.evergreenmuseum.org/exhibit/the-spruce-goose/.

Ford, "2026 F-150," https://www.ford.com/trucks/f150/?srsltid=AfmBOopyWmjrxYpdWnc040b5HFgcwqUowfAxmqFR4MZbhnEQEOC1hWmh.

Sutter, Joseph, *747: Creating the World's First Jumbo Jet and Other Adventures from a Life in Aviation* (New York: Harper Collins, 2010).

The Archeologist, "The Hellenistic Giant Galley 'Tessarakonteres,' One Of The Largest Human-Powered Vessels In History!" (February 14, 2022), https://www.thearchaeologist.org/blog/the-hellenistic-giant-galley-tessarakonteres-one-of-the-largest-human-powered-vessels-in-history.

WHY GRADUAL IMPROVEMENTS TRUMP EUREKA MOMENTS

Smil, Vaclav, *Energy and Civilization* (Cambridge, MA: MIT Press, 2017).

Smil, Vaclav, *Creating and Transforming the Twentieth Century* (New York: Oxford University Press, 2025).

Tsao, Jeff Y., "Solid-state lighting: lamps, chips, and materials for tomorrow," *IEEE Circuits and Devices Magazine* 20/3 (2004), pp. 28–37.

EUROPE'S IRREVERSIBLE DECLINE

Bonnefous, Bastien, "The great decline of European industry," *Le Monde* (September 24, 2024), https://www.lemonde.fr/en/economy/article/2024/09/23/the-great-breakdown-of-european-industry_6727000_19.html.

European Commission, "Low fertility in the EU: A review of trends and drivers," Knowledge4Policy (May 6, 2024), https://knowledge4policy.ec.europa.eu/news/low-fertility-eu-review-trends-drivers_en.

Groningen Growth and Development Centre, "Maddison Historical Statistics," University of Gronigen, https://www.rug.nl/ggdc/historicaldevelopment/maddison/?lang=en.

Noriega, Margarita, "140 years of Europe's changing population, mapped," *Vox* (June 23, 2015), https://www.vox.com/maps/2015/6/23/8824571/map-population-shifts-europe.

Index